Visual Media and Tourism

Tourism is all about visuals. Visuals stimulate our imagination, create fantasy, and drive the audiences to take actions to realize these dreams through perceived reality. With media content presented through channels of television drama, reality shows, TV commercials, and movies, this book presents findings that help us better understand the relationships between nostalgia and film tourism; how reality TV shows affect tourist experience and authenticity; and how visuals stimulate audiences' taste and olfactory senses and their relationship with gastronomical tourism. The book presents findings that explain the psychological mechanism of how modality and navigability influence tourists' behavioral intention.

With its balanced research methodology (qualitative, quantitative, and the combination of both) and important topics covered in media tourism, *Visual Media and Tourism* serves as a pertinent reference book for subjects related to special interest tourism, such as film tourism, in undergraduate programs, or modules related to research methods in both undergraduate and graduate programs. It helps readers become better informed on how visuals stimulate travel motivations, condition tourist behaviors, and affect travel experiences.

The chapters in this book were originally published as a special issue of the *Journal of Travel & Tourism Marketing.*

Seongseop (Sam) Kim is Professor, School of Hotel and Tourism Management, The Hong Kong Polytechnic University. He is a prolific author in the tourism and hospitality field. He is on the editorial board of multiple international journals.

Steve Pan is Visiting Professor, Shandong University, Weihai. His last post was Associate Professor at Singapore Institute of Technology. He served as International News Translator and Deputy Director for nearly 11 years in the News Department, China Television Company, Taiwan.

Visual Media and Tourism

Edited by
Seongseop (Sam) Kim and Steve Pan

Routledge
Taylor & Francis Group
LONDON AND NEW YORK

First published 2022
by Routledge
2 Park Square, Milton Park, Abingdon, Oxon OX14 4RN

and by Routledge
605 Third Avenue, New York, NY 10158

Routledge is an imprint of the Taylor & Francis Group, an informa business

British Library Cataloguing in Publication Data
A catalogue record for this book is available from the British Library

ISBN: 978-0-367-77013-6 (hbk)
ISBN: 978-0-367-77014-3 (pbk)
ISBN: 978-1-003-16941-3 (ebk)

Typeset in Myriad Pro
by Newgen Publishing UK

Publisher's Note
The publisher accepts responsibility for any inconsistencies that may have arisen during the conversion of this book from journal articles to book chapters, namely the inclusion of journal terminology.

Disclaimer
Every effort has been made to contact copyright holders for their permission to reprint material in this book. The publishers would be grateful to hear from any copyright holder who is not here acknowledged and will undertake to rectify any errors or omissions in future editions of this book.

Contents

Citation Information

The chapters in this book were originally published in the *Journal of Travel & Tourism Marketing*, volume 35, issue 3 (2018). When citing this material, please use the original page numbering for each article, as follows:

Preface
Preface
Seongseop (Sam) Kim and Steve Pan
Journal of Travel & Tourism Marketing, volume 35, issue 3 (2018), pp. 257–258

Chapter 1
How are food value video clips effective in promoting food tourism? Generation Y versus non–Generation Y
Seongseop (Sam) Kim, Ja Young Choe and Suna Lee
Journal of Travel & Tourism Marketing, volume 35, issue 3 (2018), pp. 377–393

Chapter 2
Flying with Nicole Kidman or Jennifer Aniston? Brand funnel stages' influence on brand personality
Filareti Kotsi and Natasa Slak Valek
Journal of Travel & Tourism Marketing, volume 35, issue 3 (2018), pp. 365–376

Chapter 3
Investigation of the technology effects of online travel media on virtual travel experience and behavioral intention
Youngjoon Choi, Benjamin Hickerson and Jinsoo Lee
Journal of Travel & Tourism Marketing, volume 35, issue 3 (2018), pp. 320–335

Chapter 4
Perceived values of TV drama, audience involvement, and behavioral intention in film tourism
Seongseop (Sam) Kim and Sangkyun Kim
Journal of Travel & Tourism Marketing, volume 35, issue 3 (2018), pp. 259–272

Chapter 5
Segmentation of potential film tourists by film nostalgia and preferred film tourism program
Seongseop (Sam) Kim and Sangkyun (Sean) Kim
Journal of Travel & Tourism Marketing, volume 35, issue 3 (2018), pp. 285–305

Chapter 6
Immersion in film tourist experiences
Yannik St-James, Jessica Darveau and Josyane Fortin
Journal of Travel & Tourism Marketing, volume 35, issue 3 (2018), pp. 273–284

Chapter 7
The role of visual media in religious tourists' destination image, choice, and on-site experience: the case of Tinos, Greece
Matina Terzidou, Dimitrios Stylidis and Konstantinos Terzidis
Journal of Travel & Tourism Marketing, volume 35, issue 3 (2018), pp. 306–319

Chapter 8
Reality television portrayals of Kavos, Greece: tourists behaving badly
Nicola Williams-Burnett, Heather Skinner and Julia Fallon
Journal of Travel & Tourism Marketing, volume 35, issue 3 (2018), pp. 336–347

Chapter 9
(Re)negotiating authenticity through virtual travel: a case study of Law of the Jungle, *a Korean reality travel program*
Sohye Kim, Sanghun Park, Carla A. Santos and Kimberly J. Shinew
Journal of Travel & Tourism Marketing, volume 35, issue 3 (2018), pp. 348–364

Notes on Contributors

Youngjoon Choi, School of Hotel and Tourism Management, The Hong Kong Polytechnic University, 17 Science Museum RD, TST East, Kowloon, Hong Kong.

Jessica Darveau, Department of Marketing, HEC Montreal, 3000 Cote-Sainte-Catherine Road, Montreal, Quebec H3T 2A7, Canada.

Julia Fallon, Cardiff School of Management, Cardiff Metropolitan University, Cardiff, UK.

Josyane Fortin, Department of Marketing, HEC Montreal, 3000 Cote-Sainte-Catherine Road, Montreal, Quebec H3T 2A7, Canada.

Benjamin Hickerson, Community and Therapeutic Recreation, The University of North Carolina at Greensboro, 206 Ferguson Bldg, 524 Highland Ave, Greensboro, NC 27402, USA.

Sangkyun Kim, Department of Tourism, School of Humanities & Creative Arts, Flinders University, Adelaide, Australia; School of Business and Law, Edith Cowan University, Joondalup WA 6027, Australia.

Seongseop (Sam) Kim, School of Hotel and Tourism Management, The Hong Kong Polytechnic University, 17 Science Road, TST East, Kowloon, Hong Kong.

Sohye Kim, Department of Recreation, Sport and Tourism, University of Illinois at Urbana-Champaign, Champaign, IL 61820, USA.

Filareti Kotsi, College of Communication and Media Sciences, Zayed University, Dubai, United Arab Emirates.

Jinsoo Lee, School of Hotel and Tourism Management, The Hong Kong Polytechnic University, 17 Science Museum RD, TST East, Kowloon, Hong Kong.

Suna Lee, School of Hotel and Tourism Management, The Hong Kong Polytechnic University, 17 Science Road, TST East, Kowloon, Hong Kong.

Steve Pan, Hospitality Business Programme, Cluster of Design and Specialised Businesses, Singapore Institute of Technology.

Sanghun Park, Department of Recreation, Sport and Tourism, University of Illinois at Urbana-Champaign, Champaign, IL 61820, USA.

Carla A. Santos, Department of Recreation, Sport and Tourism, University of Illinois at Urbana-Champaign, Champaign, IL 61820, USA.

Kimberly J. Shinew, Department of Recreation, Sport and Tourism, University of Illinois at Urbana-Champaign, Champaign, IL 61820, USA.

Heather Skinner, Institute of Place Management, Manchester Metropolitan University, Manchester, UK.

Natasa Slak Valek, College of Communication and Media Sciences, Zayed University, Abu Dhabi, United Arab Emirates.

Yannik St-James, Department of Marketing, HEC Montreal, 3000 Cote-Sainte-Catherine Road, Montreal, Quebec H3T 2A7, Canada.

Dimitrios Stylidis, The Business School, Middlesex University London, London, UK.

Konstantinos Terzidis, School of Business and Economics, Eastern Macedonia and Thrace Institute of Technology, Kavala, Greece.

Matina Terzidou, The Business School, Middlesex University London, London, UK.

Nicola Williams-Burnett, Cardiff School of Management, Cardiff Metropolitan University, Cardiff, UK.

Ja Young Choe, School of Hotel and Tourism Management, The Hong Kong Polytechnic University, 17 Science Road, TST East, Kowloon, Hong Kong.

Preface

Seeing is believing. This idiom may not be true when virtual reality and perceived reality are ubiquitous in our daily life. However, visual (and audio) content is more effective in stimulating one's affective emotions, while verbal cues are more likely to appeal to cognitive reasoning (Watson & Hill, 1993). In terms of tourism destination image (TDI) formation, the affective construct has more impact than its cognitive counterpart (Kim & Yoon, 2003). Hence, this Special Issue is dedicated to delve into the intricate and elusive relationship between visual media and its impact on tourism, be it fantasy creation, motivation/image formation, behavior conditioning, imagination congealing, or satisfaction evaluation.

As with other media, visual media comprise content and channel. In this Special Issue, two channels are mainly involved: television (TV) and films. As for the content, drama (created) and reality (gathered) are discussed. With the content and channels analyzed, contributors to this Special Issue help us better understand the value of drama/films and their effect on visit intention and experience; reality TV shows and their relationship with tourist behavior and authenticity; and food video clips in stimulating our taste and olfactory senses, which in turn help promote gastronomical tourism. Highlights of nine articles in this Special Issue are outlined in the following.

Perusing cluster analysis, Sam Kim and Sean Kim categorize film tourists into four clusters and identify nostalgia as an important segmentation variable. Nostalgia in film tourism mainly involves "reminiscence of mimicking", "memory of film backdrops", and "memory of Hong Kong history and culture". Through in-depth interviews, Yannik St-James, Jessica Darveau, and Josyane Fortin deepen and expand our understanding of how tourists manipulate and negotiate movies and TV programs to construct immersive experiences.

With an online survey of 430 Chinese respondents, Sam Kim and Sean Kim again expand our understanding of the involvement of the audience in Korean TV drama and their behavior intention by developing and testing a drama consumption model. On the other hand, Matina Terzidou, Dimitrios Stylidis, and Konstantinos Terzidis enlighten us with regard to the impact of another arguably more credible image formation agent: TV news and documentaries. Adopting participant observations and in-depth interviews, they found that broadcast images shape the image of Tinos, Greece, a religious tourism destination, and visiting patterns.

Three articles in this Special Issue delve into the topics of virtual travel, reality TV shows, and authenticity. Youngjoon Choi, Benjamin Hickerson, and Jinsoo Lee conducted an experiment with 213 participants to examine the impacts of technology affordances on virtual travel experience. Nicola Williams-Burnett, Heather Skinner, and Julia Fallon perform a qualitative thematic analysis of reality television portrayals of Kavos, Greece. The negative publicity generated by the reality TV show, according to the authors, may tarnish the image of Kavos. Sohye Kim, Sanghun Park, Carla A. Santos, and Kimberly J. Shinew, on the other hand, content-analyze a Korean reality travel program and take a closer look at authenticity, orientalism, the tourist gaze, and the second gaze.

Celebrity endorsing of products is common in advertising (Creswell, 2008). Filareti Kotsi and Natasa Slak Valek carry out research on the brand personality and its associations with two female celebrities featuring in TV commercials of two airlines in the Middle East. They found that celebrities' perceived personalities correlate with brand personality and affect consumers' decision process. The last article is related to food tourism and promotional video clips. Smell is alleged to be the sense that enables quick memory recall (Wilkie, 1994). Sam Kim, Ja Young Choe and Suna Lee analyzed the role of food value video clips which affect members of Generation Y and non-Generation Y. They compared significant differences between these two groups in terms of the perception of local food value, behavioral involvement with food featured in the clips, and intention to embark on culinary tourism at the featured destination.

Overall, nine articles in this Special Issue represent a well-balanced mix in terms of methodology (quantitative, qualitative, and a combination of both), content (film tourism, religious tourism, culinary tourism, and authenticity), and channels (movies, TV drama, TV reality shows, TV news and documentaries, TV commercials, online media, and promotional video clips). We hope and trust you will enjoy reading this Special Issue.

References

Creswell, J. (2008). Nothing sells like celebrity. *New York Times*. Retrieved from http://www.nytimes.com/2008/06/22/business/media/22celeb.html

Kim, S., & Yoon, Y. (2003). The hierarchical effects of affective and cognitive components on tourism destination image. *Journal of Travel & Tourism Marketing, 14*(2), 1–22. doi:10.1300/J073v14n02_01

Watson, J., & Hill, A. (1993). *A dictionary of communication and media studies* (3rd ed.). London: Arnold.

Wilkie, W. L. (1994). *Consumer behavior*. New York: John Wiley & Sons.

Seongseop (Sam) Kim

Steve Pan

PART 1

Visuals and travel stimulation

How are food value video clips effective in promoting food tourism? Generation Y versus non-Generation Y

Seongseop (Sam) Kim, Ja Young Choe and Suna Lee

ABSTRACT

The purpose of this study was to identify the effects of a food tourism video clip created by a destination marketing organization on potential tourists' perceptions of the destination's food values, familiarity with the destination's food, behavioral involvement with the destination's food, and intention to visit the destination for food tourism. A number of important findings were generated. First, of the five food values, "global food" and "attractive food" significantly influenced the respondents' familiarity with Hong Kong food. Second, for the Generation Y group, Hong Kong food values such as "global food", "attractive food", and "realistic restaurants" had strong effects on the viewers' behavioral involvement with Hong Kong food, while no such relationship was found between food value and behavioral involvement with Hong Kong food among the non-Generation Y group. Third, comparison of two generation groups showed that "realistic restaurants" had significantly more influence on behavioral involvement with Hong Kong food in the Generation Y group than in the non-Generation Y group. Overall, the results showed that the efficacy of the video clip in persuading potential tourists to visit the destination differed according to the generation.

Introduction

Food tourism has attracted significant research attention in the last decade (Björk & Kauppinen-Räisänen, 2016; Mkono, Markwell, & Wilson, 2013; Molz, 2007; Smith & Costello, 2009). Hall and Mitchell (2001, p. 308) gave a more rigorous definition of food tourism, such as "visitation to primary and secondary food producers, food festivals, restaurants, and specific locations for which food tasting and/or experiencing the attributes of specialist food production regions are the primary motivating factor for travel". However, a more flexible definition of food tourism is tourist activities that are totally or partially related to tasting a place's cuisine or to engaging in activities related to gastronomy (Herrera, 2012; Mak, Lumbers, & Eves, 2012).

For example, the Korea Tourism Organization promoted the country as a culinary destination in collaboration with famous celebrities, including star chefs and movie stars (Kim, Choe, & Lee, 2016). The literature indicates that many destination marketing organizations (DMOs) have transformed their major marketing tools to promote the foods of their destinations more effectively. Traditionally, DMOs have fostered food tourism through brochures, radio, and television (TV) advertisements (Frochot, 2003; Hjalager & Corigliano, 2000; Mellinger, 1994; Okumus, Okumus, & McKercher, 2007; Pritchard & Morgan, 1996). However, today, many DMOs attempt to build their brands as culinary destinations by creating video clips and posting them on social media and the Internet (Lim, Chung, & Weaver, 2012; Ontario Culinary Tourism Alliance + Skift, 2015). Although numerous DMOs have created and posted video clips on cyberspace platforms (Roque & Raposo, 2015), little is known about how food tourism video clips affect viewers' familiarity and behavioral involvement with local food and whether the video clips are effective in producing the primary desired outcomes, such as higher levels of viewers' behavioral intention to visit the destination. In particular, little research has explored the different effects of promotional video clips on different generations.

Comments, recommendations, and visual images of actual destinations on social media are considered to influence prospective tourists. Social media have become a major tool for companies in promoting certain products and communicating with customers as a marketing strategy. Lim et al. (2012, p.

199) have argued that "social media has shifted the paradigm of brand creation". In the past, information on products and advertisements used to be controlled by companies. However, today, a brand can be generated and changed by consumers who share and disseminate information about products via social media. In this regard, many DMOs have realized the value of posting video clips on social media. However, few studies have examined the efficacy of video clips in promoting destinations. As a result, the literature has failed to keep abreast of the industry trends.

The aim of this study is to examine how promotional video clips posted on social media platforms by DMOs affect viewers, particularly with respect to food tourism products. Specifically, this study has seven objectives: (1) to identify the Hong Kong food values perceived by potential tourists; (2) to assess whether the food values featured in video clips influence the viewers' familiarity with Hong Kong food; (3) to investigate whether a tourism destination's food values featured in a video clip affect behavioral involvement with the destination's food; (4) to investigate whether a tourism destination's food values featured in a video clip influence the intention to visit the destination for food tourism; (5) to explore whether familiarity with a destination's food affects the behavioral involvement with the destination's food; (6) to analyze whether familiarity and behavioral involvement with a destination's food affect the intention to visit the destination for food tourism; and (7) to examine whether two structural models differ with respect to samples of Generation Y and non-Generation Y subjects.

Food tourism in Hong Kong

In this study, Hong Kong was selected as the research site to investigate the effectiveness of food tourism video clips. There are several reasons for selecting Hong Kong as a culinary tourism destination. First, Hong Kong is a famous gastronomic tourism destination which provides unique and/or multi-ethnic cuisines, creative chefs, unique marine and agricultural products, and a unique culinary heritage (Horng & Tsai, 2012). Second, Hong Kong was ranked one of the world's best food cities in the 2014 Readers' Choice Awards (Traveler, 2016). The publication of a Michelin guide on Hong Kong has also enhanced the image of Hong Kong as an Asian culinary capital. Third, since 2009, Hong Kong has organized various culinary events, such as the Hong Kong Wine and Dine Festival, and has successfully attracted large numbers of local and international tourists.

The Hong Kong Tourism Board endeavors to promote Hong Kong's image by producing numerous promotional video clips. In 2014, it launched a global branding campaign entitled "My Time for Hong Kong" to promote the pleasurable and unique experiences that global consumers can have in Hong Kong (Hong Kong Information Services Department, 2015). Promotional video clips produced by the Hong Kong Tourism Board have won a number of renowned international awards, such as the "Marketing-Primary Government Destination" category of the 2015 PATA Gold Award and the "Campaigns" category of the 2015 ASTRID Awards. A number of food tourism video clips have been created and posted in the "My Time for Hong Kong" and "Shop and Dine" categories.

The Hong Kong Tourism Board stated that the promotional video clips were created to reflect authentic Hong Kong culture to consumers and to differentiate the destination from other tourist destinations (Hong Kong Tourism Board, 2016). For example, in one video, the three-Michelin star chef Alvin Leung described how to consume all parts of a cow in Hong Kong and introduced real Hong Kong locals. Although the real street and authentic food culture of Hong Kong may be considered unfamiliar to international tourists, it seems that the Hong Kong Tourism Board has not hesitated to show what locals actually do in real life.

Literature review

Video clips as a promotional tool for a tourism destination

Given the life cycle of a destination, emerging new competitors, and ongoing advancements in technology, a successful marketing tool to employ is continuing destination promotion (MacKay & Smith, 2006; Morgan & Pritchard, 2000; Pan, 2009; Shani, Chen, Wang, & Hua, 2010). Moreover, the characteristics of tourism products, such as the intangibility of products and the impossibility of prior testing or experience, necessitate ongoing destination promotion. A destination marketer's visual promotional activities provide effective forms of sensory stimulation that can provoke the viewer's imagination, provide virtual experiences, and convey information from experts and tourists (Chiou, Wan, & Lee, 2008; De Vries, Gensler, & Leeflang, 2012; Tussyadiah & Fesenmaier, 2009). These visual promotional tools contribute to creating favorable images and enhancing the intention to revisit (Gartner, 1994; Morgan & Pritchard, 2000; Pan, 2009).

Video clips can be readily posted on YouTube from a personal computer (PC) or other social media devices. Recently, improvements in Internet capacities, such as the speed and the amount of content possible to post, have meant that video clips can be posted in cyberspace and rapidly downloaded. YouTube is commonly used to share information among consumers and as an important marketing tool for destination marketers (Hajli, 2014; Reino & Hay, 2011). Hence, tourism organizations can maximize the effects of promoting their destinations by using new platforms such as YouTube and can decide which media type they should focus on for their destination marketing.

Recent tourism studies have shown that social media influence how tourists plan their trips (Chang & Wu, 2015; Hays, Page, & Buhalis, 2013; Hudson, Roth, Madden, & Hudson, 2015; Miguéns, Baggio, & Costa, 2008; Neff, 2012; Reino & Hay, 2011; Roque & Raposo, 2015; Tussyadiah & Fesenmaier, 2009; Xiang & Gretzel, 2010). For example, Neff (2012) found that consumers perceive a brand more positively when they watch a company's video clips on YouTube than when they view conventional advertisements such as TV commercials. Roque and Raposo (2015) compared the usefulness of social media tools among 13 DMOs and found that 12 of these DMOs preferred to use YouTube as a promotional tool, although the number of uploaded video clips differed among the DMOs. The maximum number of viewers was observed for an Australian DMO which also had the largest number of followers on YouTube. They also found that YouTube promotions maximized the effects of the DMOs' communication and marketing efforts to different degrees. Reino and Hay (2011) cited the example of Tourism Ireland's GO campaign as a case illustrating the desirability of using YouTube for destination marketing. In this campaign, short films featuring local people from Ireland were posted on YouTube. The locals introduced many Irish attractions that are not well described in the ordinary guidebooks. The local people were considered to have produced trustworthy accounts of authentic places in their hometown.

In terms of food tourism, YouTube video clips can play a greater role than other marketing approaches because food can stimulate sensory effects. For example, DMOs in South Australia produced dozens of food tourism video clips based on the theme of "South Australia's Food and Wine". In the video clips, talented local chefs, food producers, and wine makers shared their food philosophy and linked it to storytelling about the food culture of South Australia. The efforts of these DMOs were successful because Australia moved from tenth to third place behind France and Italy as one of the most preferred destinations to visit for food tourism.

However, if used incorrectly, video clips can have adverse effects for destination marketers (Reino & Hay, 2011; Zhang, 2015). For example, Reino and Hay (2011) found that outdated and unmanaged video clips on a DMO's website can make the DMO lose credibility and reduce the intention to travel. In a similar manner, Zhang (2015) observed that controversial and unreliable contents can trigger negative images of a destination in the destination choice process.

In sum, it has become a global pattern for potential tourists to choose their destinations, accommodation, and facilities/services after searching the Internet and viewing video clips that have been uploaded by tourists, businesses, and public agencies. As a result of this trend, video clips of tourism destinations have increasingly been used to promote places, enhance their brand image, and reinforce brand loyalty (Lim et al., 2012; Reino & Hay, 2011; Smith, Fischer, & Yongjian, 2012; Zhang, 2015). Although the use of video clips has become prevalent in the industry, follow-up research is only beginning to explore their effects in light of the growing worldwide trend.

Food values featured in video clips and familiarity with the food

Food has a range of values, such as taste, comfort, healthiness, exoticism, culture, and beauty (Roseman, 2006). These values can be categorized as utilitarian (e.g. the cost of food, taste of food, food portion, variety of options, and health food options), hedonic (e.g. ethnic food culture, entertainment, escape, exoticism, and ambience of the restaurant), social (e.g. social relations and interaction during meal, prestige, and status distinction by food), and epistemic (e.g. curiosity, novelty, and knowledge about food) (Adongo, Anuga, & Dayour, 2015; Chang, Kivela, & Mak, 2011; Enrique Bigné, Sánchez, & Andreu, 2009; Fields, 2002; Gyimóthy & Mykletun, 2009; Ha & Jang, 2013).

In particular, ethnic foods have a range of symbolic meanings with respect to culture, tradition, history, and ethnic authenticity. Thus, the values of food can be examined using the symbolic communications model to understand the transfer of meaning between the video clip and the viewer. According to the symbolic communications model, an advertiser strives to foster a "product meaning bundle" by linking modern cultural symbols with the product (Koernig & Boyd, 2009; Lord & Putrevu, 2009). Values, ideas, meanings, and beliefs are embedded in a product by formulating close links between the product and its symbol (Carlson & Donavan, 2008). The advertiser wants to achieve a unit association through appropriation of the symbol. As a result, the association of the symbolic icon with food elaborates its meaning bundle, which is then transmitted to current or future diners through various means of communication.

In this study, Hong Kong food is examined as a representative ethnic and regional example of Chinese cuisine. The perceptions of diverse Hong Kong food values are elicited from viewers of a YouTube video clip created by the Hong Kong Tourism Board. As shown in previous studies, a person who is exposed to a YouTube promotional video knows the precise characteristics of the product and feels closer to the product (Chang & Wu, 2015; Hudson et al., 2015; Miguéns et al., 2008; Neff, 2012; Xiang & Gretzel, 2010).

Similarly, viewers of video media on Hong Kong food are likely to form an impression because they can learn about the products and become more familiar with the cuisine. This is more probable in this study because the sample respondents had not traveled to Hong Kong before and so had not tasted the authentic cuisine in Hong Kong. Research suggests that the more people learn about Hong Kong food values through watching promotional video clips, the more likely they are to be familiar with Hong Kong food. To test this, the following hypothesis (H) is proposed:

H1: The Hong Kong food values featured in the video clip are likely to positively influence the viewer's familiarity with Hong Kong foods.

Food values featured in video clips and behavioral involvement with the food

Studies on social media have found that the destination attributes screened in video clips reinforce behavioral involvement with the destination (Tussyadiah & Fesenmaier, 2009; Wu, Wei, & Chen, 2008). For example, Tussyadiah and Fesenmaier (2009) content-analyzed 120 YouTube video clips about New York, United States (US), that were created by real tourists and 576 viewers' comments on the video clips. They found that the images presented to viewers generated pleasure and induced people to vividly imagine and become keenly interested in the destination. Wu et al. (2008) found out that Internet advertising content design significantly affected product involvement such as information search efforts and testing the product.

The video clip used in this study shows Hong Kong restaurants in which people are dining, food stands on the street, how the foods are cooked, and commentaries by food experts. The promotional video is expected to stimulate viewers' enduring involvement with Hong Kong foods, such as talking with friends about Hong Kong food, visiting Hong Kong restaurants in the viewer's country, and other related pursuits. Therefore, the second hypothesis is stated as follows:

H2: The Hong Kong food values featured in the video clip are likely to positively influence the viewer's behavioral involvement with Hong Kong foods.

Food values featured in video clips and the intention to visit the destination for food tourism

Studies have shown that memorable and impressive media created by DMOs can educate and motivate viewers and increase their intention to visit the featured place (Pan, 2009; Shani et al., 2010; Tussyadiah & Fesenmaier, 2009). Pan (2009) found that viewers of a New Zealand tourism promotional video who were satisfied and impressed with the video clip were more likely to visit the destination.

Shani et al. (2010) tested the effect of a promotional video on China as a tourism destination. After watching the video clip, the respondents perceived a more positive image of China. Among the destination attributes, "culture and nature tourism" was revealed to be the most influential factor in explaining the respondents' behavioral intention to visit China and to search for information about China as a travel destination. However, "infrastructure and superstructure" had no effect on people's future behavioral intentions. Similarly, Shao, Li, Morrison, and Wu (2016) showed the effects of a social media micro-film marketing clip created by a DMO in Shaoxing, China. After the micro-film, entitled *Cherry Blossoms for Love*, was distributed through social media, tourist numbers in the destination increased by 175.3% and the perception of the destination changed from traditional and ancient to romantic and stylish. In particular, the diverse attractions and tourism products, such as local wine, that featured in the video were found to increase viewers' intention to visit the place (Shao et al., 2016).

Overall, the literature suggests that effective tourism video clips produced by DMOs create a positive perception of a destination's attributes and are more likely to result in a stronger intention to visit the destination. Therefore, the Hong Kong cuisine depicted in the video clip is likely to enhance the viewers' intention to visit Hong Kong for food tourism. This leads to the following hypothesis:

H3: The Hong Kong food values featured in the video clip are likely to positively influence the viewer's intention to visit Hong Kong for food tourism.

Familiarity with food, behavioral involvement with food, and the intention to visit the food destination

Familiarity refers to knowledge gained through exposure to information concerning the service provider (Webb, 2000). A potential tourist can become familiar with a destination by learning about the destination's attributes through education, media, experience, and personal contact (Gursoy, 2011; Sun, Chi, & Xu, 2013; Tan & Wu, 2016). Most tourism destination studies concur that familiarity with a destination affects all destination choice procedures (Chen & Lin, 2012; Elliot, Papadopoulos, & Kim, 2011; Gursoy & McCleary, 2004; Horng, Liu, Chou, & Tsai, 2012; Lu, Chi, & Liu, 2015; Ozdemir et al., 2012; Prentice, 2004; Sharifpour, Walters, Ritchie, & Winter, 2013; Sun et al., 2013; Tan & Wu, 2016).

For example, Chen and Lin (2012) found that Chinese tourists who were more familiar with Taiwan had a positive image of the destination and a stronger intention to visit the destination. Similarly, identified that familiarity with the destination positively affected Chinese domestic tourists' satisfaction and tourists' destination loyalty through their perception of an enhanced image. Moreover, the important roles of the traditional media and online promotions were found to be conducive to raising the levels of destination familiarity among tourists. Tan and Wu (2016) found that information familiarity (degree of exposure to destination brochures, official websites, travel guidebooks, etc.) positively and directly affected Taiwanese tourists' future intention to visit Hong Kong. In addition, educational familiarity (degree of exposure to TV programs, movies, novels, and educational institutions) indirectly but still positively affected Taiwanese tourists' intention to visit Hong Kong (Tan & Wu, 2016). These results are similar to those of Elliot et al. (2011), who found that product familiarity strongly affects destination beliefs and destination receptivity. In other words, tourists who are familiar with a destination's product tend to evaluate the product more favorably and be more willing to travel to that location.

In this study, it is hypothesized that viewers' familiarity with Hong Kong foods is likely to lead them to become more behaviorally involved with these foods. Moreover, because the respondents have never visited Hong Kong before, it is reasonable to assume that the viewers become familiar with Hong Kong foods after watching the short video clip. The level of familiarity may stimulate people's curiosity to search for more information about Hong Kong foods and to become more interested in them. Thus, the following hypotheses are proposed:

H4: Familiarity with Hong Kong food is likely to positively influence behavioral involvement with Hong Kong food.

H5: Familiarity with Hong Kong food is likely to positively influence the intention to visit Hong Kong for food tourism.

Behavioral involvement with food and intention to visit the food destination

Personal involvement refers to the degree to which an individual is engaged in an activity, product, or experience (Josiam, Smeaton, & Clements, 1999; Prayag & Ryan, 2012). The level of involvement influences the intensity or amount of effort and time spent on the decision-making process (Lu et al., 2015; Prayag & Ryan, 2012). For example, Lu et al. (2015) revealed that tourists who had a higher level of involvement in diverse tourism activities were more likely to be satisfied with their trips to historic districts. In a similar vein, the level of involvement with a film was found to be closely correlated with the quality of tourists' experiences and further satisfaction with their tourism experiences on site (Kim, 2012).

A higher level of involvement with a travel product in the context of a travel agency's Internet banner advertising was found to have a significant effect on intention to purchase an air ticket (Wu et al., 2008). Similarly, the quality of visual media content has been shown to determine the level of involvement with a travel product and in turn lead to further behavioral involvement (Lim et al., 2012; Reino & Hay, 2011; Tussyadiah & Fesenmaier, 2009; Zhang, 2015). These results imply that it is important to discover which attributes of tourism video clips are more effective in inducing individuals' behavioral involvement.

In this study, behavioral involvement with Hong Kong food is mainly evaluated by viewers' interest in food, information search efforts, and communication with people who have tasted Hong Kong food or visited a Hong Kong restaurant in their homeland. Research indicates that involvement is positively related to the destination and actual visits; thus, involvement with Hong Kong food is more likely to play an important role in the behavioral intention to visit the destination. This leads to our following hypothesis:

H6: Behavioral involvement with Hong Kong food is likely to positively influence the intention to visit Hong Kong for food tourism.

Preference for social media and food tourism among Generation Y

Generation Y has been widely recognized as an important tourist market because most members of this generation already have travel experience or intend to travel abroad using the stronger spending power they have compared with other generations (Benckendorff, Moscardo, & Pendergast, 2010; Raunio, 2014). The major traits of Generation Y include more frequent travel, a strong motivation to explore new places, greater use of the Internet to book hotels and purchase tourism products, and an aspiration to receive updates (Na'Desh, 2008). This tech-savvy generation is also well known as the most visually sophisticated of any generation (Benckendorff et al., 2010).

Members of Generation Y like to share their local food experience through social media and to spend a lot of money on good food while traveling (Ontario Culinary Tourism Alliance + Skift, 2015). Destination marketers prefer to target Generation Y when they promote ethnic or regional foods (Ontario Culinary Tourism Alliance + Skift, 2015). For example, because members of Generation Y consider food tourism to be a leisure activity, the potential for the growth of wine tourism is reliant on this generation (Treloar, Hall, & Mitchell, 2004).

In sum, because members of Generation Y are more familiar with social media and tend to seek out authentic local culinary experiences, food tourism video clips are expected to have different degrees of influence on the tourist behavior of different generations. As a result, the effectiveness of food tourism video clips in capturing the attention of Generation Y and non-Generation Y viewers is expected to differ.

H7: The two structural equation models differ according to Generation Y and non-Generation Y.

Methods

Measurement

The aim of this study was to analyze the effects of a video clip on viewer perception. Thus, it was important to choose an appropriate video clip that best corresponded to the study's objectives. Two video clips developed by the Hong Kong Tourism Board in 2015 were combined to make a three-minute long video. The video clip shows the well-known British Chinese food writer and TV chef Ching-He Huang enjoying crisp pork buns and local noodles for breakfast in a local area frequented by many local people. She also introduces the ingredients of the dishes and describes the taste of the local food. Huang introduces Tim Ho Wan, the most affordable Michelin-starred restaurant in Hong Kong, and a variety of dim sum are shown. Another famous chef, Alvin Leung, then introduces various kinds of local beef dishes. Overall, the video clip shows a wide variety of local Hong Kong foods, from street food to fine dining restaurants. The video clip was originally narrated in English but was subtitled in Korean to help understanding.

On the basis of the literature review, a range of items were chosen to measure the respondents' perceptions of the food values that feature in the video clip, including the sensory and health attributes of Hong Kong food; the exotic, authentic, and vivid features of Hong Kong food; the realistic ambience of Hong Kong restaurants; and the globalized features of Hong Kong food (see for example Ha & Jang, 2013; Horng & Tsai, 2012; Kim, Kim, Agrusa, & Lee, 2012; Kivela & Crotts, 2006; Roseman, 2006). The items for describing the respondents' familiarity with Hong Kong food after watching the video clip on Hong Kong food tourism and measuring the change in their sense of closeness to or friendliness regarding Hong Kong food were adapted from previous studies (Chen & Lin, 2012; Horng et al., 2012; Lu et al., 2015).

The items measuring behavioral involvement with Hong Kong food were modified from the original food involvement scale (Bell & Marshall, 2003) and the involvement scale in the food tourism context (Getz, Robinson, Andersson, & Vujicic, 2014). The items to measure intention to visit Hong Kong for food tourism was also adapted from previous studies (Horng et al., 2012; Kim et al., 2012). These scales were measured on a 5-point Likert scale where 1 = "strongly disagree", 3 = "neutral", and 5 = "strongly agree". Gender, age, residential region, occupation, and kinds of Hong Kong foods were measured as categorical data.

Data collection

A Korean audience was selected to evaluate the video clip because Koreans have constituted one of the three major overseas national tourist groups in Hong Kong since 2010. The questions were pilot-tested using 50 Korean residents who liked Hong

Kong foods. After analyzing their comments, minor modifications were made to the original questionnaire. Data collection for the main survey was conducted from November 15 to December 5 2015 using an online survey company with 1.12 million panelists who were mostly living in large cities including Seoul, Busan, and Incheon. The main reason for adopting the online survey method was that it allows a targeted sample to be selected and data to be collected rapidly. Since this study should use experimental design such as showing a video clip on the survey website, it is most effective. Before completing the online survey, panelists received an email from the company asking whether they would participate in the survey after explaining the qualifications for being included in the sample. The promising panelists who met the qualifications were then asked to respond to two screening questions, and only those who passed the screening process of identifying the qualifications could participate in the survey.

The screening questions used to select the target sample related to age group (Generation Y member aged 20–35 years, or a non-Generation Y member aged 36–69 years) and having previously visited Hong Kong. Only people who had not traveled to Hong Kong were selected because they were more likely to give accurate assessments of the effects the promotional food video clip developed by the Hong Kong Tourism Board had on their perception of the values of Hong Kong food, familiarity with Hong Kong food, behavioral involvement, and intention to visit Hong Kong for food tourism.

The online survey platform was devised in such a way that prospective respondents answered each question of a questionnaire placed on the Internet website after viewing the short video clip as requested. Of the 1003 completed questionnaires, 967 were used for further data analysis after excluding those with insincere responses or multiple missing values.

Data analysis

First, exploratory factor analysis (EFA) was used to derive the underlying factors and confirmatory factor analysis (CFA) was used to produce the proposed measurement model, which was in turn used to specify whether a latent variable influenced an observed variable (Hair, Anderson, Tatham, & Black, 2009). Structural equation modeling (SEM) was applied to test whether the hypothesized theoretical model was consistent with the collected data for the two generational cohorts.

Results

Demographic profile and tourism activities of the respondents

The sample was divided into two groups: Generation Y, comprising the respondents aged 20–35 years (N = 430), and non-Generation Y, comprising those aged 36–69 years (N = 537). Regarding gender, there were more females (53.7%) than males (46.3%) in the Generation Y group, whereas there were slightly more males (50.3%) than females (49.7%) in the non-Generation Y group. About 82% of the Generation Y respondents were not married, while 79.5% of the non-Generation Y group were married. The Generation Y group mainly consisted of company employees (35.3%), students (30.9%), and people seeking a job (11.6%). The non-Generation Y group mainly comprised company employees (43.6%), followed by housewives (23.8%) and those who owned their own business (11.2%).

Approximately 24% of the Generation Y group had a monthly household income of less than 2 million won (US$ 1820), while 25.3% of the non-Generation Y group had a monthly household income of between 3 and 4 million won (between US$ 2730 and US$ 3640). Regarding education, about half of the Generation Y group had graduated from college (53.3%) and around a quarter were college students (27.7%). In contrast, 64.2% of the non-Generation Y group had graduated from university and 23.1% were high school graduates or below. Only 0.9% of the non-Generation Y group were university students.

Regarding the food that respondents would like to taste in Hong Kong if they visited, both the Generation Y group (80.5%) and the non-Generation Y group (78.6%) wanted to taste Hong Kong local food. While only 0.7% of the Generation Y group wanted to taste Korean food, 2.8% of the non-Generation Y group wanted to have Korean food if they visited Hong Kong. In terms of the type of restaurant that the respondents would like to visit, 22.6% of the Generation Y group and 17.5% of the non-Generation Y group wanted to visit a restaurant which is described on the Internet or in the tour guide as "a must-visit place". The most preferred type of travel mode in the case of the Generation Y group was individual + package tour (31.6%), followed by individual (59.3%) and package

Table 1. Factor analysis for the perception of food values (N = 967).

Domain	Food value perceived after watching the YouTube video clip	Factor loading	Mean
Exotic and fun (6.63[a]; 33.13[b]; .84[c])	[VL10] I thought Hong Kong foods consisted of unique food materials and ingredients compared to those of other countries.	.73	3.39
	[VL11] I felt the authentic features of Hong Kong foods.	.73.	3.50
	[VL7] I felt that Hong Kong foods were exotic.	68	3.65
	[VL8] I felt that Hong Kong foods were diverse.	.66	3.84
	[VL9] I felt that the colors of Hong Kong foods were flowery.	.65	3.53
	[VL12] I felt the tradition of Hong Kong.	.60	3.40
Global food (2.37[a]; 11.83 [b]; .81[c])	[VL14] I felt that Hong Kong foods were global.	.78	3.11
	[VL13] I felt that Hong Kong foods had international standards.	.74	3.12
	[VL16] I felt that Hong Kong foods were reliable.	.66	3.25
	[VL15] I felt that Hong Kong foods showcased the city's authentic culture.	.63	3.38
Attractive food (1.60 [a]; 8.00 [b]; .80[c])	[VL3] I felt that Hong Kong foods were attractive.	.77	3.70
	[VL2] I felt that the appearance of Hong Kong foods was good.	.77	3.68
	[VL1] I felt that Hong Kong foods were tasty.	.74	3.84
Healthy and nutritious food (1.22[a]; 6.09 [b]; .84[c])	[VL5] I felt that Hong Kong foods were healthy.	.83	3.11
	[VL4] I felt that Hong Kong foods were nutritious.	.80	3.14
	[VL6] I felt that Hong Kong foods appeared natural (e.g. ingredients are mainly natural).	.76	2.96
Realistic restaurants (1.03 [a]; 5.13[b]; .72[c])	[VL20] I felt that Hong Kong restaurants were noisy.	.75	3.30
	[VL19] I felt that Hong Kong restaurants were unsanitary and untidy.	.74	2.90
	[VL17] I felt that Hong Kong restaurants were cramped and uncomfortable.	.72	3.07
	[VL18] I felt that Hong Kong restaurants were crowded.	.72	3.40

[a]eigenvalue; [b]variance; [c]reliability alpha.

Table 2. Factor analyses for familiarity with Hong Kong food, behavioral involvement with Hong Kong food, and intention to visit Hong Kong (N = 967).

Construct	Item	Factor loading	Mean
Familiarity with Hong Kong food (2.22[a]; 74.14[b]; .82[c])	[FF2] I became friendlier after watching the video clip.	.90	3.56
	[FF1] I became familiar with Hong Kong foods after watching the video clip.	.87	3.56
	[FF3] I felt a positive image of Hong Kong foods after watching the video clip.	.81	3.54
Behavioral involvement with Hong Kong food (3.30[a]; 66.01[b]; .87[c])	[BI4] I'd like to read articles or news items concerning Hong Kong foods on a TV program or the Internet after watching the video clip.	.84	3.31
	[BI3] I'd like to search responses to tasting Hong Kong foods on the Internet after watching the video clip.	.83	3.34
	[BI5] I'd like to talk with people who tasted Hong Kong foods during their trip to Hong Kong after watching the video clip.	.81	3.47
	[BI2] I became interested in the kinds of Hong Kong foods after watching the video clip.	.80	3.63
	[BI1] I'd like to talk with people about the desire to taste Hong Kong foods after watching the video clip.	.77	3.55
Intention to visit Hong Kong (1.78[a]; 88.75[b]; .87[c])	[IVT1] I became interested in traveling to Hong Kong after watching the video clip.	.94	3.57
	[IVT2] I became interested in traveling to Hong Kong for food tourism after watching the video clip.		
		.94	3.64

[a]eigenvalue; [b]variance; [c]reliability alpha.

tour (9.1%). Meanwhile, the non-Generation Y group most preferred the individual + package tour (46.6%) mode of travel, followed by individual (29.1%) and package tour (24.4%). While 45.3% of the Generation Y group wanted to buy food and snacks, only 25.9% of the non-Generation Y group wanted to do so. Overall, there was a clear difference between the Generation Y and non-Generation Y groups in most of the areas measured.

Factor analyses and reliability tests

As shown in Tables 1 and 2, exploratory factor analyses were conducted using principal components analysis and Varimax rotation to identify the underlying dimensions and reduce the number of items for each construct. The results of the factor analysis using 20 items of food values produced five underlying domains where the eigenvalues were greater than 1.0 on the scree plot; these factors were "exotic and fun", "global food", "attractive food", "healthy and nutritious food", and "realistic restaurants".

The extracted factor structure accounted for 64.17% of the variance. The result of Barlett's test of sphericity was 8,074.82 (p = .000), indicating that one or more factors existed. The Kaiser–Meyer–Olkin (KMO) measure of sampling adequacy was .89, indicating that it was a useful validation to extract the factor structure. The factor loadings, which measured the correlation between the observed measurements and factors, were greater than Stevens' (2002) criterion (.4). The reliability alphas within the

Table 3. Confirmatory factor analysis (N = 967).

Domain or Construct	Item	Factor loading	t-value	SMC	AVE[b]	CCR[c]
Exotic and fun	VL12	.67	_a	.44	0.45	0.83
	VL11	.73	22.49	.54		
	VL10	.74	18.80	.54		
	VL9	.64	.16.60	.41		
	VL8	.63	16.43	.40		
	V7	.63	16.52	.39		
Global food	VL16	.68	_a	.47	0.49	0.79
	VL15	.71	18.81	.51		
	VL14	.72	18.73	.51		
	VL13	.68	17.83	.46		
Attractive food	VL3	.81	_a	.66	0.58	0.81
	VL2	.75	23.51	.56		
	VL1	.73	22.76	.53		
Healthy and nutritious food	VL6	.72	_a	.52	0.65	0.84
	VL5	.88	24.19	.78		
	VL4	.80	23.01	.64		
Realistic restaurants	VL18	.61	_a	.37	0.40	0.72
	VL19	.64	13.57	.41		
	VL20	.68	13.94	.47		
	VL17	.58	12.87	.34		
Familiarity with Hong Kong food	FF1	.71	_a	.50	0.56	0.79
	FF2	.74	25.57	.55		
	FF3	.80	20.41	.64		
Behavioral involvement with Hong Kong food	BI1	.76	_a	.57	0.55	0.86
	BI2	.78	23.79	.62		
	BI3	.72	21.75	.52		
	BI4	.71	21.34	.51		
	BI5	.72	21.63	.51		
Intention to visit Hong Kong for food tourism	IVT1	.88	_a	.77	0.78	0.87
	IVT2	.89	33.23	.79		
Fit indices	$\chi^2(368)$ = 994.2 (p = .000), GFI = .94, AGFI = .92, CFI = .96, NNFI = .95, RMR = .02, RMSEA = .04					

[a]In the measurement model, the estimated parameter was fixed at 1.0.
[b]Average variance extracted (AVE) = Σ(standardized loadings)^2/[(Σ(standardized loadings^2) + $\Sigma\varepsilon j$], where εj is the measurement error.
[c]Composite construct reliability (CCR) = [(Σ(standardized loadings)^2/(Σstandardized loadings)^2 + $\Sigma\varepsilon j$]
The T-values on all of the items were significant at the .001 level.

five domains were .84, .81, .80, .84, and .72; thus, the values met the .70 criterion recommended by Nunnally (1978). For this reason, the domains were considered to guarantee the internal consistency of the items in each domain. The mean values for the 20 items ranged from 2.90 to 3.84, whereas the grand means for the five domains were 3.53, 3.22, 3.74, 3.07, and 3.17 respectively.

The results of the factor analysis for familiarity with Hong Kong food, behavioral involvement with Hong Kong food, and intention to visit Hong Kong for food tourism generated a one-factor solution. Each factor explained 74.14%, 66.01%, and 88.75% of the variance respectively. The factor loadings on all of the items were greater than .77. Because all of the reliability alphas were greater than .82, the one-factor model was considered to be internally consistent for the items on a single factor. The grand means of familiarity with Hong Kong food, behavioral involvement with Hong Kong food, and intention to visit Hong Kong for food tourism were 3.55, 3.46, and 3.61 respectively.

Results of the CFAs and testing for measurement invariance between Generation Y and non-Generation Y

Before conducting the SEM procedures, CFA was used to generate the proposed measurement model, which was in turn used to indicate whether a latent variable influenced an observed variable (Kline, 1998). As shown in Table 3, the CFA results revealed a very supportive level of fit for the overall fit indices with the exception of the chi-square value ($\chi^2(368)$ = 994.2, p = .000). However, as the chi-square is sensitive to the sample size, other fit indices are substantially more helpful in evaluating the model than the chi-square value (Bollen, 1989). Goodness-of-fit analyses were then conducted for the following indices: goodness-of-fit (GFI) = .94, adjusted goodness-of-fit index (AGFI) = .92, comparative fit index (CFI) = .96, non-normed fit index (NNFI) = .95, root mean residual (RMR) = .02, and root mean square error of approximation (RMSEA) = .04.

The composite construct reliability (CCR) values were beyond the cut-off point of 0.7 (Fornell & Larcker, 1981; Hair et al., 2009); thus, each construct exhibited an acceptable level of reliability. Apart from three domains, "exotic

Table 4. Correlation (squared correlation), mean, and standard deviation.

	1	2	3	4	5	6	7	8
1 (Exotic and fun)	1.00							
2 (Global food)	.60***(.36)	1.00						
(Attractive food)	.53***(.28)	.46***(.21)	1.00					
4 (Healthy and nutritious food)	.44***(.19)	.54***(.29)	.43***(.19)	1.00				
5 (Realistic restaurants)	.13***(.02)	.09***(.01)	−.09***(.01)	−.02(.00)	1.00			
6 (Familiarity with Hong Kong food)	.50***(.25)	.50***(.25)	.65***(.42)	.46***(.21)	−.05(.00)	1.00		
7 (Behavioral involvement with Hong Kong food)	.49***(.24)	.52***(.27)	.56***(.31)	.41***(.16)	.03(.00)	.60***(.35)	1.00	
8 (Intention to visit Hong Kong for food tourism)	.53***(.28)	.49***(.24)	.61***(.37)	.39***(.15)	−.02(.00)	.64***(.41)	.66***(.44)	1.00
Mean	3.55	3.21	3.74	3.07	3.17	3.56	3.46	3.60
Standard deviation	0.56	0.56	0.58	0.60	0.53	0.59	0.62	0.72

***$p < .001$.

and fun", "global food", and "realistic restaurants", the convergent validity on the other domains was supported because the average variance extracted (AVE) values exceeded .5 (Fornell & Larcker, 1981). Since the AVE values on the three domains were a little bit lower or close to .5 (.45, 49, and .40 each) and their CCR values were higher than .7, it was considered that the convergent validity of the construct was still adequate according to Fornell and Larcker's (1981) recommendation. Furthermore, the CFA results supported convergent validity because the factor loadings for all of the instruments were significant at the .05 level (Anderson & Gerbing, 1988). To check the discriminant validity, the AVE for each domain or construct was compared with the squared correlation coefficients for the corresponding inter-constructs. For example, the AVE of "attractive food" (.58) was higher than the squared correlation coefficients (.19, .01, .42, .31, .37) for its corresponding inter-constructs, as shown in Table 4, thereby supporting discriminant validity (Fornell & Larker, 1981).

An investigation of the measurement invariance was undertaken because one of this study's objectives was to compare the results of the structural equation models for Generation Y and non-Generation Y. The multi-group SEM procedures followed the analytical procedures suggested in previous research (Byrne, 2004; Han, Back, & Barrett, 2009; Kline, 1998; Lee, Tsang, & Pan, 2015; Myers, Calantone, Page, & Taylor, 2000). A measurement invariance test was conducted to check whether the measurement model was equivalent between the two groups. The general method used to assess measurement invariance is the chi-square difference test (Kline, 1998; Myers et al., 2000). If the chi-square difference does not reveal significance in the chi-square values of two models, full metric invariance is supported, indicating that the two groups equally perceived the meanings of the measurement items (Cheung & Rensvold, 2002).

Table 5 shows a non-restricted model and the full metric variance of the model using CFA across the two groups. The chi-square difference between the non-restricted model and the full metric invariance model was 26.63, which was lower than $\chi^2_{.05}$ (22) = 33.92. It was evident that the full metric invariance model was supported, and thus the factor loadings across the two groups were assumed to be equivalent in the measurement models.

SEM

After successfully evaluating the proposed measurement model in terms of goodness of fit, reliability, validity, and measurement invariance, a structural equation model with the maximum likelihood method of estimation was used to test the hypotheses. Table 6 and Figure 1 represent the hypothesized structural model and the results of the hypotheses for the two groups.

First, for the Generation Y group, the model was not adequate as the chi-square value was statistically significant ($\chi^2(368) = 647.9$, $p = .000$). However, the other overall fit indices were excellent, with GFI = .91, AGFI = .89, CFI = .96, RMR = .03, RMSEA = .04, and NNFI = .95. Of the 18 estimated path coefficients, eight were statistically significant at the .05, .01, or .001 level. The significant relationships were found on the paths of hypothesis 1b ($\gamma_{12} = .30$, $t = 2.62$, $p < .01$), hypothesis 1c ($\gamma_{13} = .67$, $t = 9.13$, $p < .001$), hypothesis 2b ($\gamma_{22} = .36$, $t = 2.62$, $p < .01$), hypothesis 2c ($\gamma_{23} = .31$, $t = 2.14$, $p < .05$), hypothesis 2e ($\gamma_{25} = .14$, $t = 2.49$, $p < .05$), hypothesis 4 ($\beta_{21} = .37$, $t = 2.08$, $p < .05$), hypothesis

Table 5. Testing for measurement invariance across the Generation Y and non-Generation Y groups.

	Generation Y Group and non-Generation Y Group				
Model	χ^2 (df)	$\Delta\chi^2$ (df)	CFI	RMSEA	NNFI
Non-restricted model	1440.0 (736)		.95	.03	.94
Full metric invariance model (L(X)Y = IN)	1466.6 (758)	26.6 (22)	.95	.03	.94

Chi-square difference test: if $\Delta\chi^2$ (df) $<\chi^2_{.05}$ (22) = 33.92, the full metric invariance is supported.

Table 6. Structural equation modeling analysis.

Path	Unstandardized coefficient	Standard error	Standardized coefficient	*t*-value	*p*-value	Total effect	Indirect effect	Decision
Generation Y Group								
$\chi^2(368) = 647.9$ (p = .000), GFI = .91, AGFI = .89, CFI = .96, RMR = .03, RMSEA = .04, NNFI = .95								
Hypothesis 1a (γ_{11})	−.02	.10	−.02	−.20	.840	−.02	.00	Reject
Hypothesis 1b (γ_{12})	.36	.14	.30	2.62**	.009	.30	.00	Accept
Hypothesis 1c (γ_{13})	.55	.06	.67	9.13***	.000	.67	.00	Accept
Hypothesis 1d (γ_{14})	.08	.05	.08	1.48	.139	.08	.00	Reject
Hypothesis 1e (γ_{15})	−.09	.06	−.08	−1.55	.122	−.08	.00	Reject
Hypothesis 2a (γ_{21})	−.11	.12	−.10	−.92	.356	−.11	−.01	Reject
Hypothesis 2b (γ_{22})	.52	.20	.36	2.62**	.009	.47	.11	Accept
Hypothesis 2c (γ_{23})	.31	.14	.31	2.14*	.033	.55	.25	Accept
Hypothesis 2d (γ_{24})	−.11	.07	−.10	−1.71	.088	−.07	.03	Reject
Hypothesis 2e (γ_{25})	.19	.08	.14	2.49*	.013	.11	−.03	Accept
Hypothesis 3a (γ_{31})	.18	.12	.15	1.51	.131	.09	−.05	Reject
Hypothesis 3b (γ_{32})	−.11	.20	−.07	−1.05	.296	.26	.33	Reject
Hypothesis 3c (γ_{33})	.02	.15	.02	.16	.872	.55	.53	Reject
Hypothesis 3d (γ_{34})	−.07	.07	−.06	−1.05	.296	−.05	.01	Reject
Hypothesis 3e (γ_{35})	−.07	.08	−.05	−.96	.340	−.04	.01	Reject
Hypothesis 4 (β_{21})	.45	.22	.37	2.08*	.038	.37	.00	Accept
Hypothesis 5 (β_{31})	.61	.24	.46	2.56*	.011	.61	.15	Accept
Hypothesis 6 (β_{32})	.45	.09	.41	4.84***	.000	.41	.00	Accept
Non-Generation Y Group								
$\chi^2(368) = 792.1$ (p = .000), GFI = .91, AGFI = .89, CFI = .95, RMR = .02, RMSEA = .05, NNFI = .94								
Hypothesis 1a (γ_{11})	−.08	.09	−.08	−.90	.367	−.08	.00	Reject
Hypothesis 1b (γ_{12})	.24	.10	.24	2.43*	.015	.24	.00	Accept
Hypothesis 1c (γ_{13})	.52	.07	.61	7.29***	.000	.61	.00	Accept
Hypothesis 1d (γ_{14})	.14	.08	.13	1.79	.073	.13	.00	Reject
Hypothesis 1e (γ_{15})	.06	.05	.05	1.09	.275	.05	.00	Reject
Hypothesis 2a (γ_{21})	.18	.10	.16	1.81	.070	.12	−.04	Reject
Hypothesis 2b (γ_{22})	.17	.12	.14	1.44	.151	.27	.13	Reject
Hypothesis 2c (γ_{23})	.06	.10	.06	.58	.562	.38	.32	Reject
Hypothesis 2d (γ_{24})	.02	.08	.01	.21	.831	.08	.07	Reject
Hypothesis 2e (γ_{25})	−.08	.06	−.06	−1.35	.176	−.04	.03	Reject
Hypothesis 3a (γ_{12})	.21	.10	.16	2.07*	.039	.20	.03	Accept
Hypothesis 3b (γ_{31})	−.09	.12	−.07	−.82	.414	.12	.19	Reject
Hypothesis 3c (γ_{32})	.19	.10	.16	1.89	.058	.50	.34	Reject
Hypothesis 3d (γ_{33})	−.07	.09	−.05	−.87	.383	.02	.07	Reject
Hypothesis 3e (γ_{34})	.01	.06	.01	.14	.889	.03	.00	Reject
Hypothesis 4 (β_{21})	.62	.12	.52	5.06***	.000	.52	.00	Accept
Hypothesis 5 (β_{31})	.37	.13	.26	2.80**	.005	.51	.25	Accept
Hypothesis 6 (β_{32})	.55	.08	.47	6.70***	.000	.47	.00	Accept

***$p < .001$, **$p < .01$, *$p < .05$.

5 (β_{31} = .46, t = 2.56, p < .05), and hypothesis 6 (β_{32} = .41, t = 4.84, p < .001).

Second, the chi-square value was also statistically significant ($\chi^2(368)$ = 792.1, p = .000) for the non-Generation Y group. However, the other overall fit indices were acceptable, with GFI = .91, AGFI = .89, CFI = .95, RMR = .02, RMSEA = .05, and NNFI = .94. Of the 18 estimated path coefficients, six were statistically significant at the .05, .01, or .001 level. The significant relationships were found on the paths of hypothesis 1b (Y_{12} = .24, t = 2.43, p < .05), hypothesis 1c (Y_{13} = .61, t = 7.29, p < .001), hypothesis 3a (Y_{31} = .16, t = 2.07, p < .05), hypothesis 4 (β_{21} = .52, t = 5.06, p < .001), hypothesis 5 (β_{31} = .26, t = 2.80, p < .01), and hypothesis 6 (β_{32} = .47, t = 6.70, p < .001). Figure 1 shows the results of the structural modeling analysis.

Discussion and conclusion

Discussion

The purpose of this study was to identify the effects of one DMO's food tourism video clips on the viewers' perception of the destination's food values, familiarity with the destination's food, behavioral involvement with the destination's food, and intention to visit the destination for food tourism. The Hong Kong Tourism Board was selected as the DMO for this study. The results point to a number of factors that should be taken into account by DMOs when considering the production of food tourism video clips to post on social media. The most important finding was that the food tourism video clip was more effective in generating positive outcomes, such as a higher level of intention to visit Hong Kong for food tourism, for the Generation

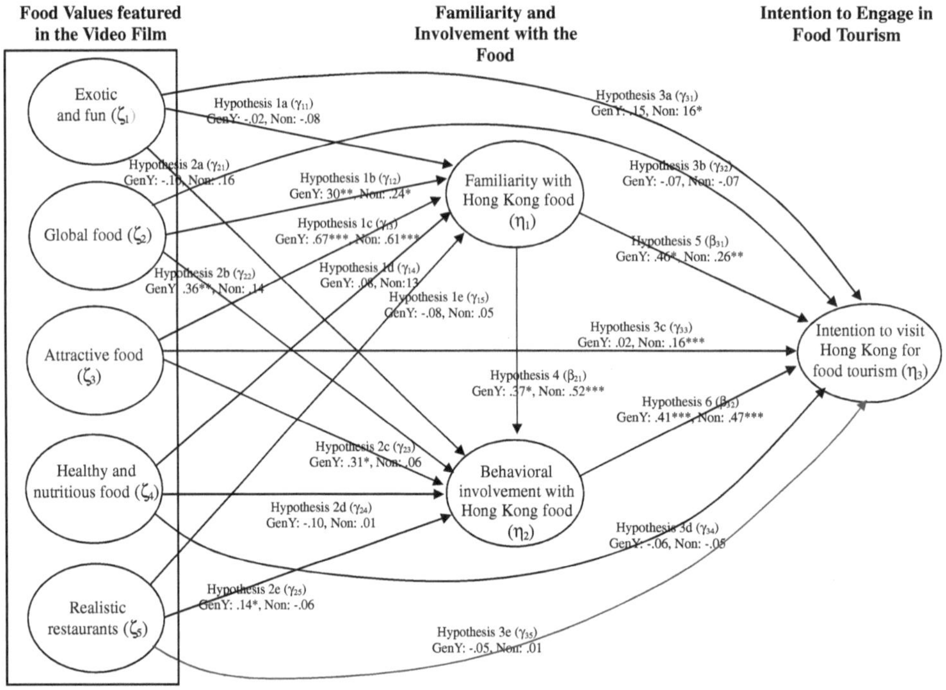

Figure 1. Results of structural model analyses.
GenY: Generation Y; Non: non-Generation Y.

Y group than for the non-Generation Y group. Also, it was found that not all of the perceptions of Hong Kong food values affected the viewers' behavioral intentions through familiarity and behavioral involvement with the food. Accordingly, DMOs need to identify the specific attributes of food values that may be effective in promoting their culinary destination.

First, the results of the SEM analyses showed that among the five values of Hong Kong food, "global food" and "attractive food" significantly influenced familiarity with Hong Kong food for both groups: that is, viewers who considered Hong Kong food to be global, international, reliable, and authentic enough to appeal to international tourists became more familiar with Hong Kong food. In addition, respondents who considered Hong Kong food to be delicious, attractive, and nicely presented became more familiar with Hong Kong food. Therefore, DMOs should strongly stress how Hong Kong food fits with international standards and how it has already been globalized to satisfy potential culinary tourists.

Also, DMOs need to emphasize the attractiveness of their foods when promoting local food in their video clips. The video clips should be designed to show and emphasize attractiveness attributes to elicit positive responses to statements such as "I felt that Hong Kong foods were global", "I felt that Hong Kong foods were reliable", "I felt that Hong Kong foods were attractive", and "I felt that the appearance of Hong Kong foods was good". Since the "global food" and "attractive food" attributes contributed to both generational groups becoming familiar with the destination's food, it is important for DMOs to bear this point in mind.

Second, for the Generation Y group, Hong Kong food values such as "global food", "attractive food", and "realistic restaurants" had strong effects on behavioral involvement with Hong Kong food. These findings support previous studies that indicated that effective YouTube video clips generate a certain level of involvement with the destination product (Reino & Hay, 2011; Tussyadiah & Fesenmaier, 2009). However, interestingly, the results for the non-Generation Y group did not support any relationship between the five values of Hong Kong food and behavioral involvement with Hong Kong food. These results showed that the video clip influenced the behavioral involvement of the two generations in different ways.

Compared to the Generation X, the Generation Y group perceived that "global food", "attractive food", and "realistic restaurants" stimulate an interest in Hong Kong food and showed more willingness to engage in activities such as actively searching for information about Hong Kong food on the Internet, dining out at a Hong Kong restaurant, and talking to people who have experienced Hong Kong food. In fact, "realistic restaurants" might be perceived either positively or negatively by viewers because the video clip shows some scenes where people queue for a long time to have a meal in a restaurant and the space of some restaurants appears cramped. Some people might not want to wait for a long time and may not enjoy a crowded environment, while others might actually prefer to go to a busy restaurant rather than an empty one. In this study, it was found that "realistic restaurants" positively affected the behavioral involvement of the Generation Y group. This indicates that local tourism-related agencies and businesses need to develop visual media that vividly portray the everyday lives of locals to differentiate Hong Kong from other destinations, especially when they target youngsters.

Third, the results show that only the "exotic and fun" value enhanced the intention to visit Hong Kong for food tourism in the case of the non-Generation Y group. This shows that most of the values of Hong Kong food perceived after watching the video clip did not directly influence these viewers' intention to visit Hong Kong for food tourism, but rather, they only influenced these viewers' intention to visit indirectly by increasing their familiarity and behavioral involvement with Hong Kong food. However, it was found that the exotic and funny aspects of the video clip had a direct influence on the non-Generation Y group's intention to visit Hong Kong for food tourism. This result indicates that the purpose of international food tourism is to seek novel food experiences rather than to taste any casual food. Therefore it implies that promotional video clips that contain unique food materials, authentic features of Hong Kong, exotic food ingredients or colors, and Hong Kong traditions could easily appeal to members of the non-Generation Y group.

Fourth, the path coefficients between familiarity with Hong Kong food and behavioral involvement were significant for both the Generation Y and non-Generation Y groups. The relationship between familiarity with Hong Kong food and intention to visit Hong Kong for food tourism was significantly effective for both groups. This supports the results of previous studies that showed that consumers' familiarity with a tourism product affects their involvement with the product and their willingness to travel to a destination (Chen & Lin, 2012; Elliot et al., 2011; Tan & Wu, 2016). It is believed that one role of video clips is to increase the viewer's familiarity with a product in the food tourism context.

Fifth, behavioral involvement with Hong Kong food was found to be significantly related to the intention to visit the destination for food tourism. This result supports a previous study that showed that a higher level of product involvement is positively related with consumers' purchase intentions (Wu et al., 2008; Xiang & Gretzel, 2010). The result further revealed that behavioral involvement with the product (Hong Kong food) can not only lead to greater consumption of the product but also enhance the intention to visit the destination for food tourism where the tourist can experience authentic and real local food.

Sixth, the results of invariance tests of paths indicated that "realistic restaurants" had significantly more influence on behavioral involvement with Hong Kong food in the Generation Y group than in the non-Generation Y group. Rather than globalizing Hong Kong food culture, it should be noted that showing the real, local, and true environments of Hong Kong restaurants is important for the Generation Y group to increase their level of behavioral involvement with Hong Kong food. The Generation Y group may prefer to watch local people in a cramped and bustling restaurant rather than tourists enjoying international food in a decent and quiet restaurant. This information can be helpful in developing video clips for future tourists.

Conclusion

The global aspects and the attractiveness of food are the most influential food values in explaining familiarity with Hong Kong food, behavioral involvement with Hong Kong food, and intention to visit Hong Kong for food tourism. This implies that the producers of video clips and audio social media should make every effort to assure consumers that Hong Kong food is trustworthy and internationally respected. Also, the media need to stimulate hedonic factors such as the packaging, appearance, and color of the foods displayed (Adongo et al., 2015; Chang et al., 2011; Ha & Jang, 2013). If the values depicted in a video clip affect the viewer's familiarity and enduring involvement with the food, the values will strongly enhance the intention to visit the destination for food tourism. Moreover, the results indicate that compared with the non-Generation Y members, members of the tech-savvy Generation Y (Na'Desh, 2008; Raunio, 2014) are more attentive to and influenced by the food values that feature in video clips and thus are more likely to demonstrate their effects in enhancing the intention to visit the food destination for food tourism through behavioral involvement with Hong Kong food.

Academic and practical implications

Few studies have examined the effects of promotional video clips on potential tourists' perceptions of destinations or their products. Research is lacking in this area because the effects of social media have only recently begun to emerge. In addition, few studies have investigated the role of video clips as promotional tools for DMOs. This study examined the influence of social media on food tourism. With the rapid advancements in information technology (IT) devices, social media are playing an increasingly important role in promoting tourism destinations. Thus, future studies need to actively analyze their influence on the mobile or IT generation to keep abreast of industry trends.

In order to study the efficacy of video clips as a promotional tool, future studies need to apply technology acceptance theories, such as the technology acceptance model (TAM) and the unified theory of acceptance and use of technology (UTAUT1, UTAUT2), which have been primarily researched in the context of the use of websites, applications, and devices in the tourism industry (Kim, Park, & Morrison, 2008; Morosan, 2011; Panagopoulos, Kanellopoulos, Karachanidis, & Konstantinidis, 2011). Moreover, future research needs to examine the efficacy of using video clips as a social media promotional tool with respect to its superiority over the traditional promotional tools, its role in social networking, reviews of consumer-generated comments, persuasiveness, and cost effectiveness.

The results of this study also have a number of practical implications for DMO marketers, local restaurants, souvenir shops, and food processing businesses. Because food incorporates a variety of non-monetary assets, such as culture, history, and tradition, and monetary assets, including income for farmers, fishermen, and businesses (restaurants, gift shops, food processors), food is an important attraction in tourism destinations (Kim et al., 2016). Thus, destination marketers need to strive to make the most of food in promoting their locations. This study also explored the efficacy of food attractiveness in appealing to potential tourists and drawing them to a destination. Furthermore, because videos are appreciated through the visual and auditory senses, it is important to consider the aesthetic elements that stimulate the senses when developing video clips.

As numerous studies have indicated, Generation Y and non-Generation Y display different information search behavior and preferences for tourism attractions (Benckendorff et al., 2010; Na'Desh, 2008; Raunio, 2014). Because the results of the comparison of the generations vary in some SEM paths, DMO marketers should consider making different video clip versions that are targeted at different markets. For example, because the Generation Y group indicated that realistic restaurant values led to a greater interest in becoming behaviorally involved with Hong Kong food, a social media promotion targeting Generation Y could be developed to emphasize the image of realistic restaurants. By applying the match-up theory between advertising endorsements and products (Kim, Wang, & Ahn, 2013; Wang, Chou, Su, & Tsai, 2007), well-known film celebrities could be employed to enhance the promotion of food to Generation Y.

Finally, the use of video clips in business-to-business (B2B), business-to-consumer (B2C), and consumer-to-consumer (C2C) communications is a worldwide trend because the clips can be readily and cost-effectively generated and distributed in cyberspace by both marketers and consumers (Reino & Hay, 2011; Wu et al., 2008; Zhang, 2015). Thus, tourism businesses and destination marketers need to actively embrace these new media platforms. In addition, tourism studies research needs to reflect the new IT trends and satisfy the needs of industry leaders to understand the efficacy of social media as a promotional tool.

Limitations and suggestions for future study

Since this study identified the effects of a video clip representing Hong Kong food on the perceptions of Korean viewers, further research is needed to compare the results of this study with those of studies which were applied to other national foods or other national tourist groups. The rationale is that the effects of food-related films can vary according to the national group (Kim, Agrusa, & Chon, 2014; Kim et al., 2012; Su, Huang, Brodowsky, & Kim, 2011). Moreover, the perception of the food featured in a visual clip can vary according to the type of food: for example, authentic food or fusion food.

The foods shown in the video clip used for this study were neither very authentic nor very global. Thus, the results of this study need to be compared to identify the distinctive effects of video clips presenting authentic foods and those presenting global foods. Further research is also needed to determine whether the results of this study differ with respect to sociodemographic variables such as gender, income, and education. Future studies need to explore the efficacy of advertising, for example, which design, promotional textual messages, and promotional model are effective.

Disclosure statement

No potential conflict of interest was reported by the authors.

Funding

The work was fully supported by a Grant from the Research Grants Council of the Hong Kong Special Administrative Region, China [Project No. PolyU156045/15B].

References

Adongo, C. A., Anuga, S. W., & Dayour, F. (2015). Will they tell others to taste? International tourists' experience of Ghanaian cuisines. *Tourism Management Perspectives, 15*, 57–64. doi:10.1016/j.tmp.2015.03.009

Anderson, J. C., & Gerbing, D. W. (1988). Structural equation modeling in practice: A review and recommended two-step approach. *Psychological Bulletin, 103*(3), 411. doi:10.1037/0033-2909.103.3.411

Bell, R., & Marshall, D. W. (2003). The construct of food involvement in behavioral research: Scale development and validation. *Appetite, 40*(3), 235–244. doi:10.1016/S0195-6663(03)00009-6

Benckendorff, P., Moscardo, G., & Pendergast, D. (Eds.). (2010). *Tourism and Generation Y*. Oxford, UK: CABI.

Björk, P., & Kauppinen-Räisänen, H. (2016). Local food: A source for destination attraction. *International Journal of Contemporary Hospitality Management, 28*(1), 177–194. doi:10.1108/IJCHM-05-2014-0214

Bollen, K. (1989). *Structural equations with latent variable*. New York, NY: John Wiley & Sons.

Byrne, B. M. (2004). Testing for multigroup invariance using AMOS graphics: A road less traveled. *Structural Equation Modeling, 11*(2), 272–300. doi:10.1207/s15328007sem1102_8

Carlson, B. D., & Donavan, D. T. (2008). Concerning the effect of athlete endorsements on brand and team-related intentions. *Sport Marketing Quarterly, 17*(3), 154–162.

Chang, H. L., & Wu, D. Y. (2015). Examining the effect of social media marketing in tourism. In *Proceedings of the 17th International Conference on Electronic Commerce 2015* (p. 4). New York: ACM.

Chang, R. C., Kivela, J., & Mak, A. H. (2011). Attributes that influence the evaluation of travel dining experience: When East meets West. *Tourism Management, 32*(2), 307–316. doi:10.1016/j.tourman.2010.02.009

Chen, -C.-C., & Lin, Y.-H. (2012). Segmenting Mainland Chinese tourists to Taiwan bydestination familiarity: A factor-cluster approach. *International Journal of Tourism Research, 14*(4), 339–352. doi:10.1002/jtr.v14.4

Cheung, G. W., & Rensvold, R. B. (2002). Evaluating goodness-of-fit indexes for testing measurement invariance. *Structural Equation Modeling, 9*(2), 233–255. doi:10.1207/S15328007SEM0902_5

Chiou, W. B., Wan, C. S., & Lee, H. Y. (2008). Virtual experience vs. brochures in the advertisement of scenic spots: How cognitive preferences and order effects influence advertising effects on consumers. *Tourism Management, 29*(1), 146–150. doi:10.1016/j.tourman.2007.03.016

De Vries, L., Gensler, S., & Leeflang, P. S. (2012). Popularity of brand posts on brand fan pages: An investigation of the effects of social media marketing. *Journal of Interactive Marketing, 26*(2), 83–91. doi:10.1016/j.intmar.2012.01.003

Elliot, S., Papadopoulos, N., & Kim, S. (2011). An integrated model of place image: Exploring relationship between destination, product, and country image. *Journal of Travel Research, 50*(5), 520–534. doi:10.1177/0047287510379161

Enrique Bigné, J., Sánchez, I., & Andreu, L. (2009). The role of variety seeking in short and long run revisit intentions in holiday destinations. *International Journal of Culture, Tourism and Hospitality Research, 3*(2), 103–115. doi:10.1108/17506180910962113

Fields, K. (2002). Demand for the gastronomy tourism product: Motivational factors. In A. M. Hjalager & G. Richards (Eds.), *Tourism and gastronomy* (pp. 36–50). London and New York: Routledge.

Fornell, C., & Larcker, D. F. (1981). Evaluating structural equation models with unobservable variables and measurement error. *Journal of Marketing Research, 18*(1), 39–50. doi:10.2307/3151312

Frochot, I. (2003). An analysis of regional positioning and its associated food images in French tourism regional brochures. *Journal of Travel & Tourism Marketing, 14*(3), 77–96. doi:10.1300/J073v14n03_05

Gartner, W. C. (1994). Image formation process. *Journal of Travel & Tourism Marketing, 2*(2–3), 191–216. doi:10.1300/J073v02n02_12

Getz, D., Robinson, R., Andersson, T. D., & Vujicic, S. (2014). *Foodies and food tourism*. Oxford, UK: Goodfellow.

Gursoy, D. (2011). Destination information search strategies. In Y. Wang & A. Pizam (Eds.), *Destination marketing and management: Theories and applications* (pp. 67–81). Wallington, UK: CABI.

Gursoy, D., & McCleary, K. W. (2004). Travelers' prior knowledge and its impact on their information search behavior. *Journal of Hospitality & Tourism Research, 28*(1), 66–94. doi:10.1177/1096348003261218

Gyimóthy, S., & Mykletun, R. J. (2009). Scary food: Commodifying culinary heritage as meal adventures in tourism. *Journal of Vacation Marketing, 15*(3), 259–273. doi:10.1177/1356766709104271

Ha, J., & Jang, S. S. (2013). Variety seeking in restaurant choice and its drivers. *International Journal of Hospitality Management, 32*, 155–168. doi:10.1016/j.ijhm.2012.05.007

Hair, J., Anderson, R., Tatham, R., & Black, W. (2009). *Multivariate data analysis* (7th ed.). Upper Saddle River, NJ: Prentice Hall.

Hajli, M. (2014). A study of the impact of social media on consumers. *International Journal of Market Research, 56*(3), 387–404. doi:10.2501/IJMR-2014-025

Hall, C. M., & Mitchell, R. (2001). Wine and food tourism. In N. Douglas., N. Douglas., & R. Derrett (Eds.), *Special interest tourism: Context and cases* (pp. 307–329). St. Milton QLD, Australia: John Wiley & Sons.

Han, H., Back, K. J., & Barrett, B. (2009). Influencing factors on restaurant customers' revisit intention: The roles of emotions and switching barriers. *International Journal of Hospitality Management, 28*(4), 563–572. doi:10.1016/j.ijhm.2009.03.005

Hays, S., Page, S. J., & Buhalis, D. (2013). Social media as a destination marketing tool: Its use by national tourism organisations. *Current Issues in Tourism, 16*(3), 211–239. doi:10.1080/13683500.2012.662215

Herrera, C. F. (2012). Gastronomy's importance in the development of tourism destination in the world. In D. Llin & I. Gaztelumendi (Eds.), *Global report on food tourism* (pp. 6–9). Madrid: UNWTO.

Hjalager, A., & Corigliano, M. A. (2000). Food for tourists? Determinants of an image. *International Journal of Tourism Research, 2*(4), 281–293. doi:10.1002/(ISSN)1522-1970

Hong Kong Information Services Department. (2015). Overseas tourism promotion [Press release]. Retrieved from http://www.info.gov.hk/gia/general/201511/18/P201511180509.htm

Hong Kong Tourism Board. (2016). Hong Kong Tourism Board social media platforms. Retrieved from http://www.discoverhongkong.com/campaign/awards/HKTBSocialMediaPlatforms.pdf

Hong Kong Circle. *Gourmet capital stars in new Michelin Guide.* Retrieved on 24 April, 2017 from http://www.hketowashington.gov.hk/dc/circle/Nov-Dec08/Articles/article09.htm

Horng, J. S., Liu, C. H., Chou, H. Y., & Tsai, C. Y. (2012). Understanding the impact of culinary brand equity and destination familiarity on travel intentions. *Tourism Management, 33*(4), 815–824. doi:10.1016/j.tourman.2011.09.004

Horng, J. S., & Tsai, C. T. S. (2012). Culinary tourism strategic development: An Asia-Pacific perspective. *International Journal of Tourism Research, 14*(1), 40–55. doi:10.1002/jtr.834

Hudson, S., Roth, M. S., Madden, T. J., & Hudson, R. (2015). The effects of social media on emotions, brand relationship quality, and word of mouth: An empirical study of music festival attendees. *Tourism Management, 47,* 68–76. doi:10.1016/j.tourman.2014.09.001

Josiam, B. M., Smeaton, G., & Clements, C. J. (1999). Involvement: Travel motivation and destination selection. *Journal of Vacation Marketing, 5*(2), 167–175. doi:10.1177/135676679900500205

Kim, D. Y., Park, J., & Morrison, A. M. (2008). A model of traveller acceptance of mobile technology. *International Journal of Tourism Research, 10*(5), 393–407. doi:10.1002/jtr.669

Kim, S. (2012). Audience involvement and film tourism experiences: Emotional places, emotional experiences. *Tourism Management, 33*(2), 387–396.

Kim, S., Agrusa, J., & Chon, K. (2014). The influence of a TV drama on visitors' perception: A cross-cultural study. *Journal of Travel & Tourism Marketing, 31*(4), 536–562. doi:10.1080/10548408.2014.883950

Kim, S., Choe, J., & Lee, A. (2016). Efforts of globalizing a national food: Segmentation by reasons for preferring Korean food. *International Journal of Contemporary Hospitality Management, 28*(10), 2010–2030. doi:10.1108/IJCHM-03-2015-0151

Kim, S., Kim, M., Agrusa, J., & Lee, A. (2012). Does a food-themed TV drama affect perceptions of national image and intention to visit a country? An empirical study of Korea TV drama. *Journal of Travel & Tourism Marketing, 29*(4), 313–326. doi:10.1080/10548408.2012.674869

Kim, S., Wang, G., & Ahn, T. (2013). Which endorser and content are most influential in Korean restaurant promotions? *International Journal of Hospitality Management, 33,* 208–218. doi:10.1016/j.ijhm.2012.08.005

Kivela, J., & Crotts, J. C. (2006). Tourism and gastronomy: Gastronomy's influence on how tourists experience a destination. *Journal of Hospitality & Tourism Research, 30*(3), 354–377. doi:10.1177/1096348006286797

Kline, R. (1998). *Principles and practice of structural equation modeling* (2nd ed.). New York, NY: The Guilford Press.

Koernig, S. K., & Boyd, T. C. (2009). To catch a tiger or let him go: The match-up effect and athlete endorsers for sport and non-sport brands. *Sport Marketing Quarterly, 18*(1), 25–37.

Lee, J. S., Tsang, N., & Pan, S. (2015). Examining the differential effects of social and economic rewards in a hotel loyalty program. *International Journal of Hospitality Management, 49,* 17–27. doi:10.1016/j.ijhm.2015.05.003

Lim, Y., Chung, Y., & Weaver, P. A. (2012). The impact of social media on destination branding: Consumer-generated videos versus destination marketer-generated videos. *Journal of Vacation Marketing, 18*(3), 197–206. doi:10.1177/1356766712449366

Lord, K., & Putrevu, S. (2009). Informational and transformational responses to celebrity endorsement. *Journal of Current Issues and Research in Advertising, 31*(1), 1–13. doi:10.1080/10641734.2009.10505253

Lu, L., Chi, C. G., & Liu, Y. (2015). Authenticity, involvement, and image: Evaluating tourist experiences at historic districts. *Tourism Management, 50,* 85–96. doi:10.1016/j.tourman.2015.01.026

MacKay, K. J., & Smith, M. C. (2006). Destination advertising: Age and format effects on memory. *Annals of Tourism Research, 33*(1), 7–24. doi:10.1016/j.annals.2005.07.002

Mak, A. H., Lumbers, M., & Eves, A. (2012). Globalisation and food consumption in tourism. *Annals of Tourism Research, 39*(1), 171–196. doi:10.1016/j.annals.2011.05.010

Mellinger, W. M. (1994). Toward a critical analysis of tourism representations. *Annals of Tourism Research, 21*(4), 756–779. doi:10.1016/0160-7383(94)90082-5

Miguéns, J., Baggio, R., & Costa, C. (2008). Social media and tourism destinations: TripAdvisor case study. *Advances in Tourism Research, 26*(28), 1–6.

Mkono, M., Markwell, K., & Wilson, E. (2013). Applying Quan and Wang's structural model of the tourist experience: A Zimbabwean netnography of food tourism. *Tourism Management Perspectives, 5,* 68–74. doi:10.1016/j.tmp.2012.10.007

Molz, J. G. (2007). Eating difference: The cosmopolitan mobilities of culinary tourism. *Space and Culture, 10*(1), 77–93. doi:10.1177/1206331206296383

Morgan, N., & Pritchard, A. (2000). *Advertising in tourism and leisure.* Oxford, UK: Routledge.

Morosan, C. (2011). Customers' adoption of biometric systems in restaurants: An extension of the technology acceptance model. *Journal of Hospitality Marketing and Management, 20*(6), 661–690. doi:10.1080/19368623.2011.570645

Myers, M. B., Calantone, R. J., Page, T. J. Jr, & Taylor, C. R. (2000). An application of multiple-group causal models in assessing cross-cultural measurement equivalence. *Journal of International Marketing, 8*(4), 108–121. doi:10.1509/jimk.8.4.108.19790

Na'Desh, F. D. (2008). *Grown up digital: Gen-Y implications for organizations* (Unpublished doctoral dissertation). Pepperdine University, Malibu, CA.

Neff, J. (2012). GE study proves consumers respond more to shared content than to paid placements. Advertising Age. Retrieved from http://adage.com/article/digital/ge-study-proves-consumers-respond-shared-content/232324/

Nunnally, J. (1978). *Psychometric theory* (2nd ed.). New York, NY: McGraw Hill.

Okumus, B., Okumus, F., & McKercher, B. (2007). Incorporating local and international cuisines in the marketing of tourism destinations: The cases of Hong Kong and Turkey. *Tourism Management, 28*(1), 253–261. doi:10.1016/j.tourman.2005.12.020

Ontario Culinary Tourism Alliance + Skift. (2015). The rise of food tourism. Retrieved from http://www.visitmarin.org/site/assets/files/3798/octa-skift-the-rise-of-culinary-tourism-2.pdf

Ozdemir, B., Aksu, A., Ehtiyar, R., Çizel, B., Çizel, R. B., & İçigen, E. T. (2012). Relationships among tourist profile, satisfaction and destination loyalty: Examining empirical evidences in Antalya region of Turkey. *Journal of Hospitality Marketing & Management*, *21*(5), 506–540. doi:10.1080/19368623.2012.626749

Pan, S. (2009). The role of TV commercial visuals in forming memorable and impressive destination images. *Journal of Travel Research*, *50*(2), 171–185. doi:10.1177/0047287509355325

Panagopoulos, A., Kanellopoulos, D., Karachanidis, I., & Konstantinidis, S. (2011). A comprehensive evaluation framework for hotel websites: The case of chain hotel websites operating in Greece. *Journal of Hospitality Marketing & Management*, *20*(7), 695–717. doi:10.1080/19368623.2011.577705

Prayag, G., & Ryan, C. (2012). Antecedents of tourists' loyalty to Mauritius: The role and influence of destination image, place attachment, personal involvement, and satisfaction. *Journal of Travel Research*, *51*(3), 342–356. doi:10.1177/0047287511410321

Prentice, R. (2004). Tourist familiarity and imagery. *Annals of Tourism Research*, *31*(4), 923–945. doi:10.1016/j.annals.2004.02.008

Pritchard, A., & Morgan, N. J. (1996). Selling the Celtic Arc to the USA: A comparative analysis of the destination brochure images used in the marketing of Ireland, Scotland and Wales. *Journal of Vacation Marketing*, *2*(4), 346–365. doi:10.1177/135676679600200406

Raunio, J. M. (2014). *Understanding the travel behavior of Generation Y* (Unpublished master's dissertation). Mid Sweden University, Sweden.

Reino, S., & Hay, B. (2011). The use of YouTube as a tourism marketing tool. In *Proceedings of the 42nd Annual Travel & Tourism Research Association Conference*, (Vol. 42). London. Travel & Tourism Research Association.

Roque, V., & Raposo, R. (2015). Social media as a communication and marketing tool in tourism: An analysis of online activities from international key player DMO. *Anatolia*, *27*(1), 1–13.

Roseman, M. G. (2006). Changing times: Consumers choice of ethnic foods when eating at restaurants. *Journal of Hospitality & Leisure Marketing*, *14*(4), 5–32. doi:10.1300/J150v14n04_02

Shani, A., Chen, P. J., Wang, Y., & Hua, N. (2010). Testing the impact of a promotional video on destination image change: Application of China as a tourism destination. *International Journal of Tourism Research*, *12*(2), 116–133.

Shao, J., Li, X., Morrison, A. M., & Wu, B. (2016). Social media micro-film marketing by Chinese destinations: The case of Shaoxing. *Tourism Management*, *54*, 439–451. doi:10.1016/j.tourman.2015.12.013

Sharifpour, M., Walters, G., Ritchie, B. W., & Winter, C. (2013). Investigating the role of prior knowledge in tourist decision making: A structural equation model of risk perceptions and information search. *Journal of Travel Research*, *53*(3), 307–322. doi:10.1177/0047287513500390

Smith, A. N., Fischer, E., & Yongjian, C. (2012). How does brand-related user-generated content differ across YouTube, Facebook, and Twitter? *Journal of Interactive Marketing*, *26*(2), 102–113. doi:10.1016/j.intmar.2012.01.002

Smith, S., & Costello, C. (2009). Culinary tourism: Satisfaction with a culinary event utilizing importance-performance grid analysis. *Journal of Vacation Marketing*, *15*(2), 99–110. doi:10.1177/1356766708100818

Stevens, J. P. (2002). *Applied multivariate statistics for the social sciences* (4th ed.). Hillsdale, NJ: Erlbaum.

Su, H. J., Huang, Y.-A., Brodowsky, G., & Kim, H. J. (2011). The impact of product placement on TV-induced tourism: Korean TV dramas and Taiwanese viewers. *Tourism Management*, *32*(4), 805–814. doi:10.1016/j.tourman.2010.06.020

Sun, X., Chi, C. G. Q., & Xu, H. (2013). Developing destination loyalty: The case of Hainan Island. *Annals of Tourism Research*, *43*, 547–577. doi:10.1016/j.annals.2013.04.006

Tan, W. K., & Wu, C. E. (2016). An investigation of the relationships among destination familiarity, destination image and future visit intention. *Journal of Destination Marketing & Management*, *5*, 214–226. doi:10.1016/j.jdmm.2015.12.008

Traveler. (2016). The best food cities in the world. Retrieved from http://www.cntraveler.com/galleries/2015-04-07/worlds-best-food-cities-readers-choice-awards-2014/13

Treloar, P., Hall, C. M., & Mitchell, R. (2004, February). *Wine tourism and the Generation Y market: Any possibilities?* Paper presented at the CAUTHE 2004 Conference: Creating Tourism Knowledge, Brisbane, Australia.

Tussyadiah, I. P., & Fesenmaier, D. R. (2009). Mediating tourist experiences: Access to places via shared videos. *Annals of Tourism Research*, *36*(1), 24–40. doi:10.1016/j.annals.2008.10.001

Wang, K. C., Chou, S. H., Su, C. J., & Tsai, H. Y. (2007). More information, stronger effectiveness? Different group package tour advertising components on web page. *Journal of Business Research*, *60*(4), 382–387. doi:10.1016/j.jbusres.2006.09.024

Webb, D. (2000). Understanding customer role and its importance in the formation of service quality expectations. *Service Industries Journal*, *20*(1), 1–21. doi:10.1080/02642060000000001

Wu, S. I., Wei, P. L., & Chen, J. H. (2008). Influential factors and relational structure of Internet banner advertising in the tourism industry. *Tourism Management*, *29*(2), 221–236. doi:10.1016/j.tourman.2007.03.020

Xiang, Z., & Gretzel, U. (2010). Role of social media in online travel information search. *Tourism Management*, *31*(2), 179–188. doi:10.1016/j.tourman.2009.02.016

Zhang, L. (2015). *The effect of social media marketing, on brand marketing and communication strategy: The case of Visit Denmark* (Unpublished master's dissertation). Roskilde University, Denmark.

Flying with Nicole Kidman or Jennifer Aniston? Brand funnel stages' influence on brand personality

Filareti Kotsi and Natasa Slak Valek

ABSTRACT

This paper reports on the visual perception of global advertising video campaigns of two United Arab Emirates (UAE) based airline companies, Etihad Airways and Emirates. Brand funnel and brand personality were tested using data from tourists travelling to UAE after the launch of television commercials featuring internationally recognized celebrities Nicole Kidman and Jennifer Aniston. The objectives of this article are to explore the relationship among television commercials' (TVCs) visuals using celebrities, brand personality, and consumers' decision-making process within the brand funnel. Findings suggest that although Etihad Airways first introduced a very important person (VIP) in its advert, Emirates emerges as a stronger player in both emirates, Abu Dhabi and Dubai. The results confirm that Etihad Airways and Emirates significantly differ in three (sincerity, excitement, competence) out of five brand personality dimensions. Empirical results of the consumers' decision process (brand funnel) in relation to brand personality dimensions are also presented.

Introduction

Etihad Airlines and Emirates launched campaigns in the same year (2015) using female celebrities that serve as brand ambassadors and opted for a strategy of catering to luxury, high-end travelers. On March 14 2015, Etihad Airlines launched its new global brand campaign "Flying Re-Imagined" with a new television commercial. It was also published on YouTube and in just over a year had more than five million views (Etihad Airways, 2015). Nicole Kidman, an Academy award-winning Australian actress, is both star and narrator of the feature, shown on global television stations and social media in both 60- and 30-second formats, with still versions of the campaign appearing in print, digital, and on outdoor channels. The television commercial (TVC) was shot onboard Etihad Airways' new flagship Airbus A380 and in various landmarks in Abu Dhabi in the United Arab Emirates (UAE), Marseille (France), and Prague (Czech Republic). The airline's chief commercial officer said: "Nicole Kidman embodies worldly sophistication, intelligence, originality, and elegance – values which form the foundations of the Etihad brand" (Etihad Airways, 2016). Seven months later, on October 5 2015, Emirates launched its television commercial, "Wake up to flying as it should be", featuring Hollywood star Jennifer Aniston. Emirates has capitalized on the American actress's approachable, comedic persona and in less than seven months it had over six million YouTube views (Emirates, 2015). The airline's Senior Vice President (SVP) for corporate communications, marketing, and brand mentioned that Jennifer Aniston's appeal and effortless connection with a global audience makes her the perfect choice for their campaign (McQueen, 2015).

With their high profile and wide recognition, celebrities are perceived as more credible than non-celebrity endorsers, exerting a greater influence on consumers' brand attitudes and purchase intent, and even though they share some common characteristics, including recognition, status, and popularity, each celebrity embodies his or her own unique image in the media (Choi & Rifon, 2007). Marketing managers base their construction of differentiation of their brands on the perception of the characteristics of brand personality (Cervera-Taulet, Schlesinger, & Yagüe-Guillen, 2013), which helps to create unique associations in the consumer memory and build brand equity (Ekinci & Hosany, 2006). The concept of brand personality, which emerged in the 1960s, defined as the set of human characteristics associated with a brand, is widely used in the field of marketing research (Cervera-Taulet et al., 2013). There is a need for empirical research that investigates the impact of brand personality on individual-level (brand awareness and purchase intent) and product-level (brand equity and market share) performance measures (Freling & Forbes, 2005; Kefallonitis, 2015). Having said that, our research problem lies in understanding the differences between two main airline companies/brands in UAE, Etihad Airways and Emirates, in terms of brand personality in relation to the brand funnel stages.

More specifically, the context in which the brand personality dimensions are researched in the present study lies within service markets, and more specifically in the airline industry. The empirical research tests the two airline companies mentioned above, surveying tourists travelling to the Middle East destination, and it compares the perception of the airlines' respective video campaigns that featured two popular internationally recognized celebrities. The purpose of the study is to explore the relationship between TVCs and video campaigns using celebrities, brand personality dimensions, and the consumers' decision-making processes within the traditional brand funnel. It aims to suggest how the brand personality dimensions influence different stages of the brand funnel, and how the use of celebrities in advertising influences the brand personality. It tests whether the two airlines can expect an increase in the number of people who can recognize and consider recommending the airlines after using celebrities in their advertisements.

Literature review

There is a limited number of studies of brand personality in the service market (Ekinci & Hosany, 2006; Murphy, Moscardo, & Benckendorff, 2007) and in the airline sector (Cervera-Taulet et al., 2013). Studies in the airline sector have mainly surveyed sponsorship, dissatisfaction, passengers' expectations, branding, brand equity, airline image, management, and communication strategies (Aksoy, Atilgan, & Akinci, 2003; Atas, Morris, & Bat, 2015; Chen & Hu, 2013; Chen & Tseng, 2010; Chong, 2007; Kefallonitis, 2015; Ko, 2016; Lim, Mohamed, Ariffin, & Guan, 2009; Lin, Huang, & Lin, 2015; Park, Robertson, & Wu, 2004; Wong & Musa, 2011).

Moreover, little is known about the influence of the brand personality dimensions with regard to driving brand loyalty and preferences (Keller & Lehmann, 2006), which should determine better brand funnel results. Brand personality becomes more important when designing differentiation strategies due to the intangible nature of services (Cervera-Taulet et al., 2013). A study in the airline business suggests that advertising influences the formation of brand personality only in the sophistication dimension (Cervera-Taulet et al., 2013). Brand personality has a positive impact on the perceived destination image and intention to recommend, when tourists ascribe personality traits to tourism destinations (Ekinci & Hosany, 2006). Park et al. (2004), in their study on Korean international air passengers, found that service value, passenger satisfaction, and airline image have a direct effect on passengers' decision-making process. Brand personality can be used to differentiate destinations and tourists articulated different destination brand personalities for each region (Murphy et al. (2007).

Excitement and sophistication were found to be associated the most strongly with perceived quality when Ramaseshan and Tsao (2007) investigated the relationship between brand personality and perceived quality testing hotels and airlines; as a result, Singapore Airlines and the Shangri-La Hotel were personified by descriptors such as "upper class" and "charming". Also, Chong (2007), in his study on Singapore Airlines, supported the idea that the most visible aspect of the airlines' projected identity is the "Singapore Girl". Evidently, media have played a major role in the iconic embodiment of the "Singapore Girl" brand. Similarly, our research focuses on the impact of celebrities on the airlines' brand personality, which in the later stage, according to our hypothesis, has a final impact on the consumers' decision-making process.

Celebrities and media

Magazine advertisements (47.2%), newspapers (42.3%), and television (42.8%) are still leading the consumer's first exposure to brands, initiating the consumer decision-making process (Belch & Belch, 2014). Media play an important role in consumer product consumption and TVCs are one of the channels used to form desirable destination images; however, tourism advertising, especially tourism TVCs, is rarely researched (Pan, 2011). Touristic experience is multisensory and involves more than just the visual; however, the influence of the "gaze" remains strong (Pan & Ryan, 2009). Under the assumption that "what is beautiful is good", advertisers often choose celebrities based on their physical attractiveness, intending to gain from the dual effects of celebrity status and physical appeal (Friedman & Friedman, 1979; Singer, 1983, as cited in Kamins, 1990). Absent from existing literature, which focuses almost exclusively on developed Western countries, is how celebrities might be used in the Arab world (Abdulla & Keenan, 2015). Advertisers are aware of the positive influence that celebrities can bring to a persuasive message and approximately 25% of all United

States (US) TVCs feature celebrities (Choi & Rifon, 2007) used to maintain attention and create high recall (Malik & Sudhakar, 2014).

The effectiveness of celebrities is underpinned by the public's fascination with sports personalities and movie stars, but they need to be carefully selected, matching the target segment and brand values with the features of the celebrity (Seneviratne & Molesworth, 2015). The typical consumer may symbolically aspire to identify with this group by purchasing the product recommended by the celebrity (Assael, 1984, as cited in Kamins, 1990). If a celebrity is recognized, recall of information endorsed by this celebrity is superior to when an individual fails to recognize the celebrity. (Molesworth, Seneviratne, & Burgess, 2015). The results of a study on the effectiveness of using creative marketing techniques such as celebrities and humour to improve passenger attention to the pre-flight safety briefing revealed the humorous pre-flight safety briefing video to be the most effective, as well as aiding in the recall of key safety messages (Seneviratne & Molesworth, 2015).

Although several other researchers were interested in brand management proposing symbolic components (Keller, 1993; Ligas, 2000), only a few models implementing symbolic brand meaning were validated. Aaker's (1997) model is one of the few validated and also defines the brand personality construct at the individual brand level. Thus, it is a logical consequence to implement Aaker's (1997) brand personality model into the brand funnel levels. However, there is a lack of empirical evidence related to the impact of brand personality on brand awareness and purchase intent through the use of celebrities in TVCs, and therefore this research aims to contribute to this growing body of literature.

Research hypotheses

As mentioned earlier, the aim of the present study is to understand the influence of brand personality dimensions on the brand funnel stages. The idea is inspired by social identification theory (Tajfel & Turner, 1979). It is argued that individuals will have the tendency to achieve and maintain a positive self-distinctiveness, which will be highly motivated by membership in a group. In other words, the social identification theory relates to a person's sense of belonging to a certain group or organization (Kim, Han, & Park, 2001; Mael & Ashforth, 1992), which in our case is the consumer's identification with the brand (Etihad Airways or Emirates) or the differentiation of the brand from others. Our assumption is that consumers' identification with the brand and brand funnel stages will be different depending on whether they have or have not experienced the brand. Thus, our hypothesis is that by narrowing the brand funnel, less significant differences or no differences at all between people who have and people who have not had an experience with the provided brand (airlines companies in our case) will be found. By studying previous research in this same context, we believe that the more a brand is experienced by people, which is tested by the brand funnel, the more the brand personality should be unified. This means that the latest brand stages should show fewer or less pronounced differences in brand personality dimensions between people who have an experience with the tested brand and those who have not had this same experience.

Method

Instrument

A questionnaire was developed consisting of three main sections: (1) brand awareness, (2) decision making-process (from consideration to word-of-mouth (WOM)), and (3) brand personality. One questionnaire was developed, but was adjusted to each airline company. Consequently, data were collected separately, in Dubai, testing Emirates, and in Abu Dhabi, testing Etihad Airlines. Emirates' home base is in Dubai, whereas Etihad is based in Abu Dhabi. In the first part of the questionnaire (1), top-of-mind awareness of the airline company was researched among tourists travelling to Dubai and Abu Dhabi, followed by the recognition question.

The study used an adapted model from Aaker's (1997) theoretical framework of the brand personality construct and investigated its antecedents and consequences (Eisend & Stockburger-Sauer, 2013). The advertising campaigns' celebrity endorsement is used to test antecedents that create brand personality while the consequences (recognition, consideration, purchase, loyalty, WOM) are tested using the traditional brand funnel. Aaker's (1997) seminal work on the theoretical framework of the brand personality construct that focuses on the symbolic use of brands and the understanding of how and when brand personality relates to a consumer's personality and influences consumer preference, has inspired many of the studies that validated or refined the brand personality scale (BPS) across contexts and cultures. Geuens, Weijters, and De Wulf (2009) developed a new brand personality measure consisting of five factors (activity, responsibility, aggressiveness, simplicity and emotionality), but this has not yet been applied or validated.

Aaker's (1997) 42 personality traits that additionally support the stability of the five personality dimensions (sincerity, excitement, competence, sophistication, and ruggedness) have served as a brand personality measure in several studies (Cervera-Taulet et al., 2013; Eisend & Stockburger-Sauer, 2013; Ekinci & Hosany, 2006; Freling & Forbes, 2005; Murphy et al., 2007; Ramaseshan & Tsao, 2007). Aaker's brand awareness pyramid inclusive of recall and recognition brand awareness measurement, was used at the beginning of our self-administered questionnaire (as cited in Cooper, Fletcher, Fyall, Gilbert, & Wanhill, 2008). Pictures of celebrities featured in global video campaigns, Nicole Kidman for Etihad Airlines and Jennifer Aniston for Emirates, were tested for recognition (Figures 1 and 2). The image-based visual test for video awareness has been shown to be successful in previous studies (Balasubramanian, Jagannath, & Adalarasu, 2013; Khan & Khan, 2012; Pan, 2011) and has been used since Aaker and Bruzzone (1981) claimed that the consumers' recognition of visual excerpts from television advertisement is the minimum sufficient evidence proving the advertisement "really worked".

In the following section of the questionnaire (2), respondents evaluated statements on a 5-point Likert scale (1 – strongly disagree, 5 – strongly agree), testing the consideration, purchase intention, loyalty, and WOM for each airline separately. It was previously proposed by researchers to use a scale design when rating the product (Austin, Siguaw, & Mattila, 2003). The scale can also be used when trying to differentiate respondents for segmentation purposes based on the ratings of the products, which is consistent with the aim of our research. Thus, a scale was used in the present research, which was checked by Cronbach's alpha for reliability and consistence of the questionnaire ($\alpha = 0.63$).

The final part of the survey (3) consisted of five traits describing brand personality as proposed by Aaker (1997), but modified for our research. To have any practical value, the scale should be short and easy to administer since brand personality is often only one of several measures in a questionnaire (Geuens et al., 2009). Murphy, Moscardo, Benckendorff (2007) also did not adopt the entire list of Aaker's (1997) 42 personality traits because of the risk of respondent fatigue. Thus, for each brand dimension (human personality structure as defined by Norman, 1963) one trait was chosen, but tested as a dichotomous nominal variable. The tested variables are presented in Table 1.

For the demographic background section, three variables were included: age, gender, and country of origin.

Figure 1. Emirates' global advertisement image featuring Jennifer Aniston.
Source: Emirates (2015).

Figure 2. Etihad Airways' global advertisement image featuring Nicole Kidman.
Source: Etihad Airways (2015).

Data collection and sample

Data were collected separately in Dubai (testing Emirates) and Abu Dhabi (testing Etihad Airlines). The aim was to collect data from tourists visiting the two emirates, Abu Dhabi and Dubai. To get tourists' data, the survey was conducted in-person at the most popular tourism attractions in the two emirates: Sheikh Zayed Grand Mosque was chosen in Abu Dhabi and Dubai Mall in the emirate of Dubai. Visitors' statistical data show that the Sheikh Zayed Grand Mosque is the most visited attraction by tourists visiting Abu Dhabi: 4.2 million visitors were recorded in 2013 (Sheikh Zayed Grand Mosque Statistics, 2013). On the other hand,

Table 1. Brand personality dimensions and dichotomous variables tested.

Brand personality dimension (Norman, 1963)	Trait (Aaker, 1997)	Opposite variable (defined by authors)
Sincerity	Family-oriented	Individualistic
Excitement	Trendy	Classic
Competence	Confident	Shy
Sophistication	Glamorous	Plain
Ruggedness	Masculine	Feminine

Dubai Mall is a nine million square foot shopping center, the largest mall in the world when built (El-Adly, 2007), welcoming 75 million visitors in 2013 (Nagraj, 2014). It is located adjacent to the world's tallest building, Burj Khalifa, a "must see" attraction in Dubai.

The questions were administered in the English language, which is the most appropriate language for international tourists visiting UAE. Data were collected in the same period, February and March 2016, in both emirates. The convenience sampling method was used, but potential participants were asked a screening question before they were deemed eligible to participate. Only those who identified themselves as international tourists were selected to participate in this research. This was needed because expatriates represent 87% of the UAE population, so tourists had to be identified (World Population Review, 2016). The questionnaire took approximately 8–10 minutes to complete. Finally, 140 questionnaires were collected in Abu Dhabi and 170 in Dubai. Samples by age groups and gender are presented in Table 2. Due to a high variety in nationalities visiting UAE the first seven nationalities according to the highest share are presented in Table 3, and those with N < 6 are summed into the category "Other". The sample was self-selected, meaning that statistically sample bias applies, but it is believed that the sample reasonably well represents the population of tourists in both emirates, based on the gender and nationality representation (45 nationalities are included in the sample of Abu Dhabi tourists and 36 different nationalities are included in the sample of Dubai tourists).

Table 2. Abu Dhabi and Dubai sample by gender and age groups.

Sociodemographic variable		Abu Dhabi N = 140 N	%	Dubai N = 170 N	%
Age (years)	18–25	31	22.1	49	28.8
	26–35	50	35.7	63	37.1
	36–45	31	22.1	37	21.8
	46–60	28	20.0	21	12.4
Gender	Male	48	34.3	82	48.2
	Female	87	62.1	78	45.9
	Missing	5	3.6	10	5.9

Table 3. Samples by tourists' nationalities.

Abu Dhabi sample Nationality	N	%	Dubai sample N	%	Nationality
India	17	12.1	38	22.4	Saudi Arabia
USA	12	8.6	16	9.4	Qatar
UK	10	7.1	14	8.2	Egypt
Kuwait	9	6.4	10	5.9	Kuwait
Germany	9	6.4	9	5.3	USA
China	8	5.7	8	4.7	Germany
Philippines	8	5.7	6	3.5	India
Others	67	47.9	69	40.6	Other
Total	140	100.0%	170	100.0%	**Total**

Data analysis

Descriptive statistics were calculated for all parts of both (Etihad Airlines, Abu Dhabi and Emirates, Dubai) questionnaires, inclusive of demographic variables. Two marketing funnels were developed, for each airline company separately. In the following, both databases were merged into one, which made it possible to understand the top-of-mind airlines' recall in both emirates together, Abu Dhabi and Dubai. Questions about purchase, loyalty, and WOM were specific to the tested brand (e.g. Loyalty: "I would buy Etihad Airlines/Emirates even though another airline were to offer a special price for the same itinerary"). Furthermore, purchase, loyalty, and WOM were tested on a 5-point Likert scale and only the highest agreement with the statement (5 – totally agree) was taken into consideration in the marketing funnels. Despite the fact that "4 – agree" and "5 – totally agree" could be summed together and presented as (better) results, our interest was in those tourists who were indeed the most convinced, would purchase, be loyal, and recommend the airline tested.

Continuing with the last part of the questionnaire, merged data of brand personality dimensions for Etihad Airlines and Emirates were compared. The T-test for brand personality dimensions was run, including Levene's test for equality of variances, to understand the differences in brand dimensions for the two airlines. Finally, the influence of each marketing funnel stage on each brand personality dimension was investigated. An independent T-test was used for testing the differences, separately for each airline company. The calculations for brand personality difference which have been performed are the following:

- Recognition: answers of tourists who did recognize the advert with the celebrities (recognition "yes") and those who did not (recognition "no") were compared.
- Consideration: answers of tourists who considered Emirates/Etihad Airlines for their next purchase (consideration "yes") and those who consider other airlines (consideration "no") were compared.
- Purchase: tourists who totally agreed (Likert value M = 5) to purchase Emirates/Etihad Airlines for their next purchase (purchase "yes") and all other values (Likert value M < 5; purchase "no") were compared.
- Loyalty: tourists who totally agreed (Likert value M = 5) to buy Emirates/Etihad Airlines flight tickets

even if another airline were to offer special prices for the same itinerary (loyalty "yes") and all other values (Likert value M < 5; loyalty "no") were compared.

- WOM: tourists who always recommend Emirates/ Etihad Airlines (Likert value M = 5) to their family and friends (recommendation "yes") and all other values (Likert value M < 5, recommendation "no") were compared.

Every questionnaire used (Emirates and Etihad Airlines) was independently tested for validation and reliability. The total variance result for the Emirates questionnaire is 28.556% of total variance and 25.131% of total variance for Etihad Airways. Total variance for the single factor is acceptable if lower than 50%, which can be confirmed in our case.

Results

Marketing funnel results for Etihad Airlines and Emirates are presented in Figure 3. Top-of-mind recall and recognition techniques were used to test the awareness. Etihad Airlines is at the top of the mind of 21.4% of tourists travelling to Abu Dhabi, while Emirates is at the top of the mind of 29.4% of tourists travelling to Dubai. The recognition results show an evident difference: 24.4% of tourists in Abu Dhabi could recognize the picture taken from the advertisement with Nicole Kidman; on the other hand, the picture from the Emirates advertisement with Jennifer Aniston was recognized by more than half (53.7%) of the tourists travelling to Dubai.

Moreover, all brand funnel stages for Emirates are stronger when compared with Etihad Airlines: consideration 26.7% for Etihad Airlines and 50.1% for Emirates, purchase intention 26.7% for Etihad Airlines and 46% for Emirates. The lowest value can be found in the loyalty stage, with 13.3% for Etihad Airlines and 16.1% for Emirates. Also, the biggest drop from the previous stage can be noticed in Emirates loyalty (from 46% to 16.1%). Etihad Airlines loyalty is similarly low, but the drop is not so drastic (from 26.7% to 13.3%). More tourists would recommend Emirates (41.6%), followed by Etihad Airlines (34.8%).

Finally, two stages of the brand funnel for UAE (Dubai and Abu Dhabi) are also presented in Figure 3 showing that Emirates is a top-of-mind airline for 28.4% of tourists travelling to Dubai or Abu Dhabi. Half of the tourists (49.8%) travelling to Dubai or Abu Dhabi consider Emirates as the most probable brand for purchasing a ticket from if they had to book a flight now. Although Qatar Airways was not researched in the presented study, it is worth mentioning that 9.5% of respondents in a merged data set recalled Qatar Airways, and 17.1% would consider Qatar Airways the most if they were to book a flight now. Thus, Qatar Airways' consideration in UAE is higher than the consideration of Etihad Airlines, but lower than Emirates' consideration.

Brand personality results for Emirates and for Etihad Airlines are presented in Figure 4.

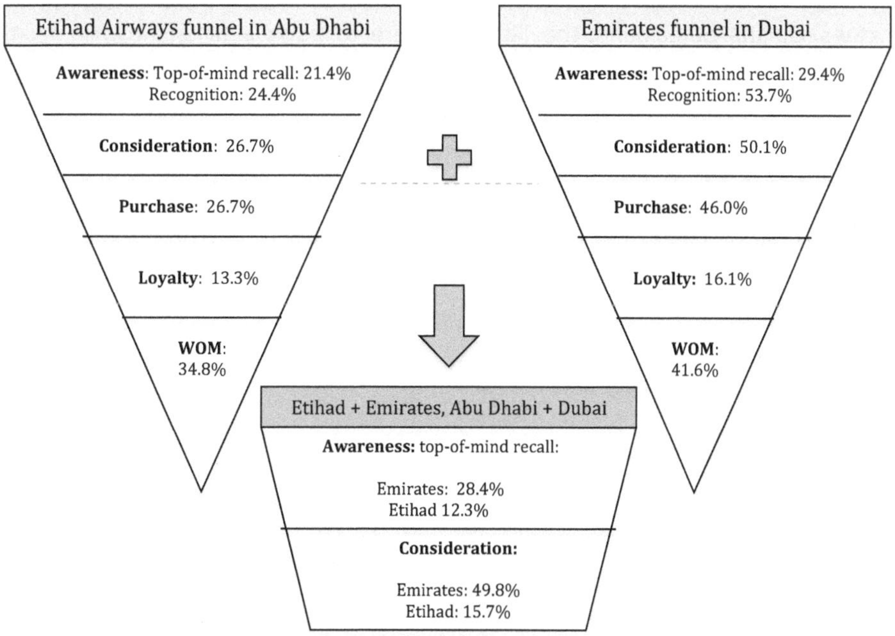

Figure 3. Airlines marketing funnels in the United Arab Emirates.
WOM: word-of-mouth.

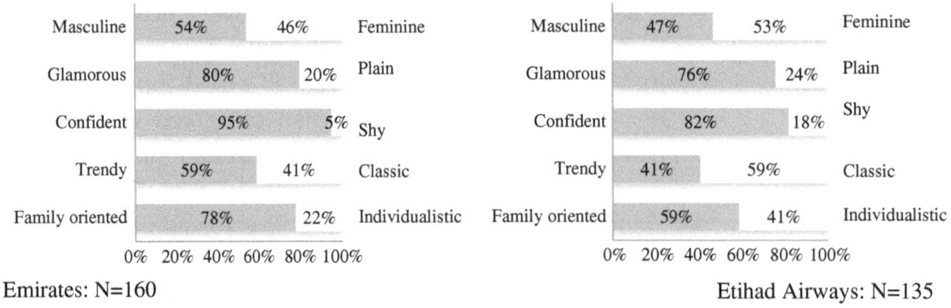

Figure 4. Brand personality for Emirates and Etihad Airways.

Based on the results from our respondents, namely international tourists travelling to Dubai, Emirates' personality is seen as more masculine (54%) than feminine (46%), much more glamorous (80%) than plain (20%), very much more confident (95%) than shy (5%), trendier (59%) than classic (41%), and more family-oriented (78%) than individualistic (22%). On the other hand, 53% of respondents in Abu Dhabi see Etihad Airlines as more feminine (53%) than masculine (47%) and more classic (59%) than trendy (41%), which are the two brand personality dimensions where Etihad Airlines and Emirates are seen differently.

In order to confirm the differences in brand personality dimensions, the independent *t*-test was performed. The results show significant differences (at the 0.05 level, marked in bold in Table 4) in three of five tested brand personality dimensions: sincerity ($t(262.759) = 3.518$, $p = 0.001$), excitement ($t(293.000) = 3.122$, $p = 0.002$) and competence ($t(204.518) = 3.428$, $p = 0.001$). The Levene's test for equality of variance is also significant in those same three cases. On the other hand, the T-test does not show significant differences in sophistication and ruggedness when comparing Emirates and Etihad Airlines brands.

Finally, the T-test for brand personality dimensions and five marketing-funnel stages was performed. The results are presented in Table 5, where the significant *p*-value at the level 0.05 is marked in bold with the *t*-statistics in parenthesis. The *t*-test did not show significant differences in brand personality dimensions for tourists that were able to recognize Emirates/Etihad Airlines advertisement photos with Jennifer Aniston/Nicole Kidman (marked as "recognition 'yes'" in Table 5) and those who did not (sig > 0.05).

The next stage, consideration, is significant when related to sincerity ($p = 0.018$, $t = -2.416$) and competence ($p = 0.004$, $t = -2.918$) in Etihad Airlines' case. Tourists who would consider Etihad Airlines for an immediate ticket see the airline as more family-oriented (M = 1.25, SD = 0.439) while tourists who consider other airlines see Etihad as more individualistic (M = 1.46, SD = 0.501). Also, tourists considering Etihad Airlines see this company as confident (M = 1.06, SD = 0.232), whereas tourists considering other airline companies see Etihad Airlines as more shy than confident (M = 1.22, SD = 0.418). On the other hand, no significant differences were found between tourists who would and those who would not consider Emirates, if they were to book a ticket now, in relation to any brand personality dimensions (sig>0.05).

The later funnel stages show the following results: tourists who are willing to purchase an Emirates ticket in the future and those who are not see Emirates differently in terms of sincerity ($p = 0.019$, $t = 2.371$), competence ($p = 0.004$, $t = -2.951$), and ruggedness ($p = 0.019$, $t = -2.362$), while tourists willing to purchase

Table 4. T-test for brand personality of Etihad airways and Emirates.

		Airline	N	Mean	SD	*p*	*t*
Sincerity	Family-oriented – Individualistic	Etihad	135	1.41	0.493	**0.001***	3.518
		Emirates	160	1.22	0.415		
Excitement	Trendy – Classic	Etihad	135	1.59	0.493	**0.002***	3.122
		Emirates	160	1.41	0.494		
Competence	Confident – Shy	Etihad	135	1.18	0.384	**0.001***	3.428
		Emirates	160	1.05	0.219		
Sophistication	Glamorous – Plain	Etihad	135	1.24	0.431	0.364	0.910
		Emirates	160	1.20	0.401		
Ruggedness	Masculine – Feminine	Etihad	135	1.53	0.501	0.188	1.319
		Emirates	160	1.46	0.500		

Marked in **bold** = significant at 0.05 level.
*Levene's test for equality of variances significant (F sig.<0.05, $t > 3$).

Table 5. T-test for marketing funnel stages and brand personality dimensions for Emirates and Etihad Airways.

	Recognition		Consideration		Purchase		Loyalty		WOM	
	no (other)	yes (Emirates)	no (other)	yes (Emirates)	no (M < 5)	yes (M = 5)	no (M < 5)	yes (M = 5)	no (M < 5)	yes (M = 5)
Emirates	N = 75	N = 85	N = 80	N = 80	N = 87	N = 73	N = 94	N = 66	N = 94	N = 66
	sig. (*t*)		sig. (*t*)		sig. (*t*)		sig. (*t*)		sig. (*t*)	
Family-oriented – Individualistic	0.175 (−1.364)		0.086 (1.726)		**0.019** (−2.371)		**0.000** (−3.995)		**0.027** (−2.227)	
Trendy – Classic	0.984 (−0.020)		0.750 (0.319)		0.214 (1.248)		0.063 (1.870)		0.471 (−0.722)	
Confident – Shy	0.194 (1.306)		1.000 (0.000)		**0.004** (−2.951)		0.494 (0.685)		0.059 (−1.904)	
Glamorous – Plain	0.694 (−0.394)		0.238 (1.184)		0.063 (−1.870)		**0.027** (−2.282)		0.081 (−1.757)	
Masculine – Feminine	0.380 (−0.881)		0.875 (0.158)		**0.019** (−2.362)		**0.007** (−2.841)		0.319 (−1.000)	
	no (other)	yes (Etihad A.)	no (other)	yes (Etihad A.)	no (M < 5)	yes (M = 5)	no (M < 5)	yes (M = 5)	no (M < 5)	yes (M = 5)
Etihad Airways	N = 102	N = 33	N = 99	N = 36	N = 99	N = 36	N = 117	N = 18	N = 88	N = 47
	sig. (t)		sig. (t)		sig. (t)		sig. (t)		sig. (t)	
Family-oriented – Individualistic	0.301 (−1.038)		**0.018** (−2.416)		0.139 (−1.499)		**0.001** (−3.835)		**0.000 (−5**.296)	
Trendy – Classic	0.530 (0.630)		0.794 (0.262)		0.087 (−1.723)		0.865 (0.171)		0.676 (0.419)	
Confident – Shy	0.653 (0.451)		**0.004** (−2.918)		**0.004** (−2.918)		**0.000** (−5.471)		**0.003** (−3.003)	
Glamorous – Plain	0.311 (1.022)		0.056 (−1.943)		**0.001** (−3.336)		0.086 (−1.778)		**0.000** (−3.003)	
Masculine – Feminine	0.874 (−0.159)		0.215 (−1.246)		0.486 (−0.698)		0.074 (−1.872)		0.738 (−3.003)	

Marked in **bold** = significant at 0.05 level; WOM: word-of-mouth.

an Etihad Airlines ticket in the future and those who will not purchase an Etihad Airlines ticket in the future statistically differ in competence (p = 0.004, t = −2.918) and sophistication (p = 0.001, t = −3.336). Tourists who are loyal to Emirates significantly differ in three main dimensions in comparison with disloyal tourists: sincerity (p = 0.000, t = −3.995), sophistication (p = 0.027, t = −2.282), and ruggedness (p = 0.007, t = −2.841). Tourists loyal to Etihad Airlines on the other hand differ from tourists not loyal to Etihad in sincerity (p = 0.001, t = −3.835) and competence (p = 0.000, t = −5.471). Finally, the recommendation (WOM) as the lowest funnel stage affects the sincerity in both cases (p = 0.027, t = −2.227 for Emirates and p = 0.000, t = −5.296 for Etihad Airlines). The sophistication (p = 0.003, t = −3.003) and ruggedness (p = 0.000, t = −3.003) is significant for Etihad Airlines results while for Emirates it is not ($p > 0.05$).

Discussion and conclusions

The findings of this study seem to indicate that Emirates is the strongest airline in both emirates, Dubai and Abu Dhabi. Although the emirate of Abu Dhabi has its own national carrier and was also the first to use an international celebrity for a global promotion, Emirates still appeared to be a much stronger airline in the minds of people travelling to UAE. Emirates results appear to be much stronger in comparison to Etihad Airlines in all stages of the marketing funnel, although Etihad Airlines released a global video marketing campaign first (April 2015). Also, after merging results of both emirates, Emirates appeared stronger again, which indicates that a later start of advertising (October 2015) was a good strategy and it could also be a reason why Emirates appear to be much stronger in peoples' mind than Etihad, especially in terms of recognition (56% for Emirates and 24% for Etihad Airlines). It should be noted that some of the differences in basic recognition and strength of the airlines may be due to the fact that Emirates was founded in 1985 and Etihad was not founded until 2003.

Looking at the brand consideration, the results coincide with the identification of world airlines. According to World Airline Awards (2015), also described as "the Oscars of the aviation industry", which announces every year the top 100 airlines in the world, Emirates ranked fifth in 2015, losing one place compared with 2014 (ranked fourth), followed by Etihad Airlines (sixth in 2015), who, by contrast, gained three places in one year (ninth in 2014). The top is reserved for Qatar Airways, which is also the biggest competitor in the UAE market as confirmed by our research. Qatar Airways' consideration was in fact stronger (17.1%) than Etihad Airways' (15.7%) according to our results, which can be considered a plus for UAE's national carrier. Etihad Airlines are advised to track and observe Qatar Airways transporting tourists to UAE. Emirates leads in "consideration" in Dubai as well as in the UAE with 50%.

Besides consideration, the purchase intention of Emirates was also close to 50%. Pfeiffer and Zinnbauer (2010) claimed that online advertising drives activity particularly in the later funnel stages, which generate actual revenues. Thus, Emirates' and Etihad Airlines' strategies to advertise on television and online at the same time with the video featuring celebrities seems to be a success. Although purchase intention for Etihad Airlines was much lower (26.7%) than for Emirates (46%), it is still in line with the previous funnel stages results (recognition 24.4% and consideration 26.7%), meaning that a quarter of people who consider Etihad Airlines would indeed also purchase a ticket with Etihad Airlines. Loyalty for Etihad appears to be slightly lower (13.3%) compared with the previous stage, while Emirates loses significantly in loyalty (16.1%) compared with the previous stage (46%). Yoon and Uysal (2005) confirmed that loyalty is positively affected by tourist satisfaction with their experiences, thus research into customer satisfaction should be considered. Also, Emirates are advised that there is a need to work more on loyalty programs or find other strategies for making their customers more loyal.

The last funnel stage (WOM) again shows better results for both companies, indicating that people with an Etihad Airlines/Emirates experience do recommend this same company. Since satisfaction and recommendation are positively correlated (Lee, Yoon, & Lee, 2007) and emotional values have a big influence on satisfaction (Sweeney & Soutar, 2001) an airline brand also needs to work on emotions. Since an affective image is the main antecedent of loyalty (Hernández-Lobato, Solis-Radilla, Moliner-Tena, & Sánchez-García, 2006), a stronger understanding on how both brands, Emirates and Etihad Airlines, are seen by their customers, in terms of brand personality, should be beneficial. Brand personality is not easy to create, but once created it tends to have a long life (Kim et al., 2001) and this is why it needs to be strategically planned and studied before. Beyond such identification, this study, within the framework of social identification theory, determines brand funnel stages influence on brand personality. Moreover, Freling and Forbes (2005) claim that there is a clear need for empirical research that investigates the impact of brand personality associated with brand awareness and purchase intent, which was carried out in the presented research.

Since two popular personalities were featured in the recent promotion of both airlines, brand personality results might reflect the celebrity. In fact, Emirates' incorporation of Jennifer Aniston seemed to create a more masculine and trendy personality, while Etihad Airlines' use of Nicole Kidman is seen as having a more feminine and classic personality. Since Etihad Airlines' TVC went with a more traditional airline advertisement style, "Flying Re-imagined" and the Oscar-winning ambassador, trying to convince higher-demand travellers, the result in brand personality was expected. On the other hand, Emirates used a more comedic fan-based TVC, which resulted in a trendier brand. Cervera-Taulet et al. (2013) found that advertising only influences sophistication among the five dimensions of brand personality and this was confirmed for an airline brand in Spain. Supporting our results, where sophistication was found to be a strong dimension (much more glamorous than plain) for both tested airlines with no significant differences, it could be concluded that Jennifer Aniston, as well as Nicole Kidman, have added glamour to the advertisements. On the other hand, it is also crucial for the airline industry to understand the importance of differentiation (Ekinci & Hosany, 2006) especially when appearing in the same market(s), which was successfully implemented by Emirates and Etihad Airlines. They differ in three (sincerity, excitement, competence) out of five brand dimensions tested.

Taking into consideration our final finding, it can be suggested that advertisement recognition does not have a significant effect on any of the brand personality dimensions in our case. It can be said that neither Nicole Kidman nor Jennifer Aniston seemed to influence how UAE airlines were seen in terms of brand personality. Consideration, the next funnel stage, brings different results for Etihad Airlines and Emirates. No effect on even one brand personality dimension in Emirates' case was found, therefore consideration is important for the Etihad Airlines brand in two dimensions: sincerity and competence. People who would consider Etihad if they were to book a ticket now, see this airline as more family-oriented and confident than those who would not consider Etihad. Thus, in order to increase consideration, Etihad Airlines should work more on a family-oriented image and a confident brand.

The latter funnel stages seem to be more vulnerable to brand personality dimensions. It might be because, as observed by Briggs, Krishnan, and Borin (2005, p. 82), "the lower funnel is about converting market shoppers into buyers". In both cases, it can be confirmed that tourists in our samples who have or do not have a purchase intention (purchase "yes"/purchase "no") significantly differ in seeing airlines' sophistication ($p = 0.004$ in both cases). Thus, it can be concluded that tourists would consider purchasing a flight ticket with a more confident rather than a shy company, and therefore reject Khan & Khan's (2012) finding which states that the attitude towards an advertisement does not determine the purchase intention at all. Moreover, our findings also indicate in both cases that tourists are more loyal to an airline and recommend it when it is seen to be more family-oriented than individualistic. Those final results are important findings for the airlines whose global campaigns aim is to increase the number of loyal shoppers (purchase and loyalty) and to be recommended (WOM).

An important overall conclusion can be drawn by our research. It was suggested that the personality of a brand enters into consideration in later funnel stages. When people become familiar with a brand (awareness) they all tend to see the brand in the same way as the same personality, whereas in the later funnel stages, at the purchase and loyalty stages in particular, it becomes crucial to work on a personality that can provide better selling results for the brand, which seems to be true for both Emirates and Etihad Airlines.

Limitations and future research

Our research has some limitations that should be addressed in terms of future research. As the traditional marketing funnel could not be elaborated as a whole for the Dubai and Abu Dhabi merged data, further research allowing this is recommended. Also, a "before–after" study is recommended to test the direct influence of celebrities in airline advertisements and the impact they leave on the airline brand personality. Using an internationally recognized person in a global campaign is understandable, but since the campaign with Nicole Kidman attracted criticism as an anti-female advert (which was published in a notable number of media worldwide), a study in this context would be beneficial before creating a new campaign.

Disclosure statement

No potential conflict of interest was reported by the authors.

References

Aaker, D. A., & Bruzzone, D. E. (1981). Viewer perceptions of prime-time television advertising. *Journal of Advertising Research, 21*(5), 15–23.

Aaker, J. (1997). Dimensions of brand personality. *Journal of Marketing Research*, *34*, 347–356. doi:10.2307/3151897

Abdulla, R., & Keenan, K. L. (2015). Perceptions of celebrity advertising in Egypt: A focus on gender. *Journal of Communication & Media Research*, *7*(1), 8–14.

Aksoy, S., Atilgan, E., & Akinci, S. (2003). Airline services marketing by domestic and foreign firms: Differences from the customers' viewpoint. *Journal of Air Transport Management*, *9*, 343–351. doi:10.1016/S0969-6997(03)00034-6

Atas, U., Morris, G., & Bat, M. (2015). The importance of sponsorship in corporate communication: A case study of Turkish Airlines. *Global Media Journal*, *6*(11), 35–53.

Austin, J. R., Siguaw, J. A., & Mattila, A. S. (2003). A re-examination of the generalizability of the Aaker brand personality measurement framework. *Journal of Strategic Marketing*, *11*(2), 77–92. doi:10.1080/0965254032000104469

Balasubramanian, V., Jagannath, M., & Adalarasu, K. (2013). EEG-based evaluation of viewer's response towards TV commercials. *International Journal of Industrial and Systems Engineering*, *13*(4), 480–495. doi:10.1504/IJISE.2013.052611

Belch, G., & Belch, M. (2014). The role of new and traditional and media in the rapidly changing marketing communications environment. *International Journal on Strategic Innovative Marketing*, *1*, 130–136.

Briggs, R., Krishnan, R., & Borin, N. (2005). Integrated multichannel communication strategies: Evaluating the return on marketing objectives- the case of the 2004 Ford F-150 launch. *Journal of Interactive Marketing*, *19*(3), 81–90. doi:10.1002/dir.20045

Cervera-Taulet, A., Schlesinger, M. W., & Yagüe-Guillen, M. J. (2013). Influence of advertising on brand personality in the airline sector: The case of Spain. *Journal of Travel & Tourism Marketing*, *30*(5), 445–454. doi:10.1080/10548408.2013.803390

Cooper, C., Fletcher, J., Fyall, A., Gilbert, D., & Wanhill, S. (2008). Tourism. *Principles and practice*. Harlow: Prentice Hall.

Chen, C., & Tseng, W. (2010). Exploring customer-based airline brand equity: Evidence from Taiwan. *Transportation Journal*, *49*(1), 24–34.

Chen, P., & Hu, H. (2013). The mediating role of relational benefit between service quality and customer loyalty in airline industry. *Total Quality Management & Business Excellence*, *24*(9/10), 1084–1095. doi:10.1080/14783363.2012.661130

Choi, S., & Rifon, N. J. (2007). Who is the celebrity in advertising? Understanding dimensions of celebrity images. *The Journal of Popular Culture*, *40*(2), 304–324. doi:10.1111/jpcu.2007.40.issue-2

Chong, M. (2007). The role of internal communication and training in infusing corporate values and delivering brand promise: Singapore Airlines' experience. *Corporate Reputation Review*, *10*(3), 201–212. doi:10.1057/palgrave.crr.1550051

Eisend, M., & Stockburger-Sauer, N. E. (2013). Brand personality: A meta-analysis review of antecedents and consequences. *Marketing Letters*, *24*(3), 205–216. doi:10.1007/s11002-013-9232-7

Ekinci, Y., & Hosany, S. (2006). Destination personality to tourism destinations. *Journal of Travel Research*, *45*, 127–139. doi:10.1177/0047287506291603

El-Adly, I. M. (2007). Shopping malls attractiveness: A segmentation approach. *International Journal of Retail & Distribution Management*, *35*(11), 936–950. doi:10.1108/09590550710828245

Freling, T. H., & Forbes, L. P. (2005). An empirical analysis of the brand personality effect. *Journal of Product and Brand Management*, *14*(7), 404–413. doi:10.1108/10610420510633350

Geuens, M., Weijters, B., & De Wulf, K. (2009). A new measure of brand personality. *International Journal of Research in Marketing*, *26*(2), 97–107. doi:10.1016/j.ijresmar.2008.12.002

Hernández-Lobato, L., Solis-Radilla, M. M., Moliner-Tena, M. A., & Sánchez-García, J. (2006). Tourism destination image, satisfaction and loyalty: A study in Ixtapa-Zihuatanejo, Mexico. *Tourism Geographies*, *8*(4), 343–358. doi:10.1080/14616680600922039

Kamins, M. A. (1990). An investigation into the 'match-up' hypothesis in celebrity advertising: When beauty may be only skin deep. *Journal of Advertising*, *19*(1), 4–13. doi:10.1080/00913367.1990.10673175

Kefallonitis, E. (2015). Airline brand dissatisfaction: An overview. *International Journal on Strategic Innovative Marketing*, *2*, 1–16.

Keller, K. (1993). Conceptualizing, measuring, and managing customer-based brand equity. *Journal of Marketing*, *57*, 1–22. doi:10.2307/1252054

Keller, K. L., & Lehmann, D. R. (2006). Brands and branding: Research findings and future priorities. *Marketing Science*, *25*(6), 740–759. doi:10.1287/mksc.1050.0153

Khan, B. M., & Khan, S. (2012). Attitude towards advertising and brand preference in Indian context-An empirical assessment. *BVIMR Management Edge*, *5*(2), 81–87.

Kim, C. K., Han, D., & Park, S. B. (2001). The effect of brand personality and brand identification on brand loyalty: Applying the theory of social identification. *Japanese Psychological Research*, *43*(4), 195–206. doi:10.1111/1468-5884.00177

Ko, Y. D. (2016). An airline's management strategies in a competitive air transport market. *Journal of Air Transport Management*, *50*, 53–61. doi:10.1016/j.jairtraman.2015.10.005

Lim, K. Y., Mohamed, R., Ariffin, A., & Guan, G. G. (2009). Branding an airline: A case study of Air Asia. *Malaysian Journal of Media Studies*, *11*(1), 35–48.

Lee, C. K., Yoon, Y. S., & Lee, S. K. (2007). Investigating the relationships among perceived value, satisfaction, and recommendations: The case of the Korean DMZ. *Tourism Management*, *28*(1), 204–214. doi:10.1016/j.tourman.2005.12.017

Ligas, M. (2000). People, products and pursuits: Exploring the relationship between consumer goals and product meanings. *Psychology and Marketing*, *17*(983), 1003. doi:10.1002/1520-6793(200011)17:11<983::AID-MAR4>3.0.CO;2-J

Lin, A. Y., Huang, Y., & Lin, M. (2015). Customer-based brand equity: The evidence from Taiwan. *Contemporary Management Research*, *11*(1), 75–94. doi:10.7903/cmr.14153

Mael, F. B., & Ashforth, E. (1992). Alumni and their alma mater: A partial test of the reformulated model of organizational identification. *Journal of Organizational Behavior*, *13*, 103–123. doi:10.1002/(ISSN)1099-1379

Malik, A., & Sudhakar, B. (2014). Brand positioning through celebrity endorsement. A review contribution to brand literature. *International Review of Management & Marketing*, *4*(4), 259–275.

Molesworth, B. R. C., Seneviratne, D., & Burgess, M. (2015). Selling safety: The use of celebrities in improving awareness of safety in commercial aviation. *Ergonomics*, 1–6. doi:10.1080/00140139.2015.1109712

Murphy, L., Moscardo, G., & Benckendorff, P. (2007). Using brand personality to differentiate regional tourism destinations. *Journal of Travel Research*, *46*, 5–14. doi:10.1177/0047287507302371

Norman, W. T. (1963). Toward an adequate taxonomy of personality attributes: Replicated factor 54 structure in peer nomination personality ratings. *Journal of Abnormal and Social Psychology*, *66*, 574–583. doi:10.1037/h0040291

Pan, S. (2011). The role of TV commercial visuals in forming memorable and impressive destination images. *Journal of Travel Research*, *50*(2), 171–185. doi:10.1177/0047287509355325

Pan, S., & Ryan, C. (2009). Tourism sense-making: The role of the senses and travel journalism. *Journal of Travel & Tourism Marketing*, *26*(7), 625–639. doi:10.1080/10548400903276897

Park, J.-W., Robertson, R., & Wu, C.-L. (2004). The effect of airline service quality on passengers' behavioural intentions: A Korean case study. *Journal of Air Transport Management*, *10*, 435–439. doi:10.1016/j.jairtraman.2004.06.001

Pfeiffer, M., & Zinnbauer, M. (2010). Can old media enhance new media? *Journal of Advertising Research*, *50*(1), 42–49. doi:10.2501/S0021849910091166

Ramaseshan, B., & Tsao, H.-Y. (2007). Moderating effects of the brand concept on the relationship between brand personality and perceived quality. *Brand Management*, *14*(6), 458–466. doi:10.1057/palgrave.bm.2550090

Seneviratne, D., & Molesworth, B. R. C. (2015). Employing humour and celebrities to manipulate passengers' attention to pre-flight safety briefing videos in commercial aviation. *Safety Science*, *75*, 130–135. doi:10.1016/j.ssci.2015.01.006

Sweeney, J. C., & Soutar, G. N. (2001). Consumer perceived value: The development of a multiple item scale. *Journal of Retailing*, *77*(2), 203–220. doi:10.1016/S0022-4359(01)00041-0

Tajfel, H., & Turner, J. C. (1979). An integrative theory of intergroup conflict. *The Social Psychology of Intergroup Relations*, *33*(47), 74.

Wong, K. M., & Musa, G. (2011). Branding satisfaction in airline industry: A comparative study of Malaysia Airlines and Air Asia. *African Journal of Business Management*, *5*(8), 3410–3423.

Yoon, Y., & Uysal, M. (2005). An examination of the effects of motivation and satisfaction on destination loyalty: A structural model. *Tourism Management*, *26*(1), 45–56. doi:10.1016/j.tourman.2003.08.016

Web references

Emirates. (2015, October 5). Emirates A380 featuring Jenifer Aniston. Retrieved May 17, 2016, from https://www.youtube.com/watch?v=kwYr4LAIUjk

Etihad Airways. (2015, March 14). Flying re-imagined. TV commercial featuring Nicole Kidman. Retrieved May 17, 2016, from https://www.youtube.com/watch?v=sXlmF3el9R0

Etihad Airways. (2016). Retrieved May 10, 2016, from http://www.etihad.com/en-us/experience-etihad

McQueen, A.-M. (2015, October 5). Emirates releases first look at light-hearted Jennifer Aniston advert. *The National*. Retrieved May 11, 2016, from http://www.thenational.ae/arts-lifestyle/celebrities/20151005/8550/emirates-releases-first-look-at-lighthearted-jennifer-aniston-advert

Nagraj, A. (2014, January 30). Dubai Mall welcomes 75 million visitors in 2013. *Gulf News*. Retrieved April, 2017, from http://gulfbusiness.com/dubai-mall-welcomes-75m-visitors-in-2013-up-15-from-2012/

Sheikh Zayed Grand Mosque Statistics. (2013). Retrieved May 5, 2016, from http://www.szgmc.ae/en/open-data

World Airlines Awards. (2015). Retrieved May 6, 2016, from http://www.worldairlineawards.com/awards/world_airline_rating.html

World Population Review. (2016). Retrieved June 3, 2016, from http://worldpopulationreview.com/countries/united-arab-emirates-population/

Investigation of the technology effects of online travel media on virtual travel experience and behavioral intention

Youngjoon Choi, Benjamin Hickerson and Jinsoo Lee

ABSTRACT

This study investigates technology affordances of online travel media determining ways of presenting pictorial information and creating the spatial structure of a destination: modality and navigability. To examine the impacts of technology affordances on virtual travel experience, a 2 (modality: still pictures versus panoramic pictures) × 2 (navigability: absence versus presence) between-subjects experiment was conducted with 213 participants. This study found significant effects of modality and navigability on affective and cognitive dimensions of virtual travel experience. Scrutinizing the mediating role of virtual travel experience, the findings explain the psychological mechanism of how modality and navigability influence tourists' behavioral intention.

Introduction

With technology innovation, online travel media have transformed and diversified ways of presenting pictorial information about a tourist destination. For instance, pictorial information is enriched and reshaped in a virtual environment where tourists can explore and navigate a hyperlinked and multiple-layered web space. As media technology enables tourists to search high-quality and navigable pictorial information, online travel information search is considered to elicit virtual travel experience (Cho, Wang, & Fesenmaier, 2002; Hyun & O'Keefe, 2012). Virtual travel experience refers to the mediated experience of exploring and appreciating the cultural and spatial landscapes of a destination. By adopting advanced media technology (e.g. high-resolution visual images, panoramic pictures, interactive and navigable functions, and three-dimensional (3D) virtual reality), many travel websites have developed various web interfaces to improve the quality of virtual travel experience (Cho et al., 2002; Chung, Han, & Joun, 2015; Jung, Chung, & Leue, 2015).

Each web interface is composed of different types and levels of affordance, which refers to "a particular capability possessed by the medium to facilitate a certain action" (Sundar, 2009, p. 556). Distinguishing the effects of a medium (e.g. technical features of a web interface) and a message (e.g. the content of pictorial information), researchers in media effect studies argue that technology affordances embedded in a web interface have the potential to induce substantial psychological effects on media users' perceptions and behaviors (Reeves & Nass, 1996; Sundar, 2008). According to Sundar (2008), each technology affordance may produce different heuristics (e.g. realism, being there, elaboration, play) influencing the quality of the user experience. Likewise, not only the content of pictorial information, but also the technical aspect of online travel media can determine the quality of the virtual travel experience (Hyun & O'Keefe, 2012; Wu & Chang, 2005).

Conceptualizing virtual travel as "a combination of the geography of the environment and user motion" (Balakrishnan & Sundar, 2011, p. 168), there are two salient technology affordances determining the way of presenting pictorial information of a destination and creating the spatial structure of a virtual environment: modality and navigability. Modality, a technology affordance determining the mode of communication (Sundar, 2008), is related to the way of presenting pictorial information and visual spatiality of a destination (e.g. still picture and panoramic picture). Navigability refers to an affordance determining the degree of user movement control in a virtual environment, such as clicking or dragging to move from point A to point B (2008). For travel websites, modality and navigability are regarded as main technology affordances influencing the quality of the virtual experience.

From the perspective of destination marketing organizations (DMOs) and tourism marketers, an important question is how to create an optimal user interface that produces a positive virtual travel experience and increases behavioral intention to visit and recommend to others. Considering the potential psychological effects produced by a user interface, it is important to understand the relationship between technology affordances and virtual travel experience. Also, because each user interface is composed with various technology affordances (Reeves & Nass, 1996; Sundar, 2008), it is necessary to distinguish different types of technology affordance and understand corresponding effects.

Although two research streams have been developed to understand the impact of online media, there exist several research gaps. One stream comes from media effect studies identifying and articulating the distinctive psychological effects of affordances on user experience (Reeves & Nass, 1996; Sundar, 2008). However, media effect studies have paid less attention to examining the effects of affordance on tourists' experience and behavior. The other stream is from tourism studies supporting that online information search influences the formation of a destination image (Horng & Tsai, 2010; Jeong, Holland, Jun, & Gibson, 2012; Lepp, Gibson, & Lane, 2011) and tourists' decision-making process (Cormany & Baloglu, 2011; Kah & Lee, 2014; López-Bonilla & López-Bonilla, 2013; Michaelidou, Siamagka, Moraes, & Micevski, 2013). However, these tourism studies do not distinguish the effects of technology (medium) and content (message) by adopting an object-centered approach, not a variable-centered approach. Because the object-centered approach treats the multiple attributes of a certain medium as a whole (e.g. TV versus Internet), it is not possible to distinguish the unique effect of each technology affordance. In a sense that online media have been developed as a heterogeneous form of multimedia (Reeves & Nass, 1996; Sundar, 2008), it is critically important to adopt the variable-centered approach to understand the underlying mechanism of how modality and navigability induce unique effects on tourists' perception and behavioral intention.

More importantly, given that the combination of two or more technology affordances may possibly magnify or offset certain psychological effects (Sundar, 2008; Sundar, Xu, & Dou, 2012), the variable-centered approach enables the detection of the interaction effect of modality and navigability. For instance, the effect of navigability can be stronger or weaker in one particular presentation mode than the other. There has been no study examining the interaction effect of modality and navigability.

In order to fill the research gaps, the purpose of this paper is (1) to examine the psychological effects of modality and navigability on virtual travel experience, and (2) to explain the psychological mechanism of how modality and navigability influence tourists' behavioral intention. Explaining human-computer interactions in the context of virtual travel, media equation theory and media richness theory would provide a fundamental theoretical lens to understand the technology effects of online travel media to enhance the quality of virtual travel experience. The investigation of technology effect on virtual travel experience would articulate key technology attributes of online travel media and examine its substantial psychological effects.

Literature review

In order to understand the effects of technology affordances on virtual travel experience and behavioral intention, this study first conceptualizes the multidimensionality of virtual travel experience. Next, this study proposes the distinctive psychological effect of modality and navigability on the affective and cognitive dimensions of virtual travel experience. Then, the relationship between virtual travel experience and behavioral intention is established.

Affective and cognitive dimensions of virtual travel experience

Traditionally, touristic experience is considered to have both affective and cognitive elements of temporality (Jennings & Weiler, 2006), a tourist moment (Cary, 2004), and the tourist gaze (Urry, 2002). In terms of affect, touristic experience contains positive and negative feelings of being outside of the ordinary daily life and confronting different cultures. At the same time, tourists are consistently involved in cognitive processes of collecting and recognizing markers (a piece of information about the space) to organize their experience (MacCannell, 1989). As symbols or spatial cues, markers can provide tourists with information about the structure of a destination. Similarly, in a virtual environment, tourists can experience emotional and cognitive responses to the mediated experience when exploring, being transported to, and feeling the virtual destination (Jun & Vogt, 2013; Tang & Jang, 2012). In this regard, virtual travel experience is conceptualized as a multifaceted concept reflecting positive and negative experiential attributes.

Capturing both affective and cognitive dimensions, this study conceptualizes virtual travel experience with spatial self-location, affective engagement, disorientation, and a spatial mental model. Three dimensions (spatial self-location, affective engagement, and disorientation) are related to affection, and one (spatial mental model) represents cognition. Related to a sense of being there, spatial self-location refers to an individual's feeling of being part of the mediated environment and forgetting about the immediate space (Balakrishnan & Sundar, 2011; Cho et al., 2002). Affective engagement means an individual's emotional attachment to the moment of exploring a destination in a virtual environment (Flowerday & Schraw, 2003). Referring to a feeling of distraction or dizziness, disorientation reflects the negative affection when exploring a virtual environment (Zhang, Appelman, Choi, & Han, 2013). A spatial mental model means a tourist's level of understanding of the spatial structure of a virtual environment by recognizing markers that convey the information of the destination (Balakrishnan & Sundar, 2011).

Impact of modality on virtual travel experience

Modality refers to an affordance shaping the mode of communication (input and output of information) or the presentation of a message, such as text, audio, picture, and video (Sundar, 2000). Urry (2002) posits that tourists are primarily dependent on their vision when experiencing the landscapes of a destination. Hence, although travel media have been developed as a multimedia platform, it is not surprising that the presentation mode of pictorial information plays a fundamental role in conveying the experiential quality of a destination (MacKay & Fesenmaier, 1997; Wearing, Stevenson, & Young, 2010). Ever since photography was introduced in the 19th century, there have been various ways of presenting pictorial information (Jacobs, 2004). Recently, online travel media have been offering new platforms to (re)construct pictorial information in a virtual environment (Kim & Richardson, 2003; Tussyadiah & Fesenmaier, 2009). Whereas traditional travel media present still pictures as the primary presentation mode in magazines, brochures, and postcards, online travel media are using seamlessly integrated visuals to portray the landscapes of a destination (e.g. virtual reality simulations, 3D virtual environments, 360° panoramic pictures, and video clips).

For travel websites, one recent interesting aspect of modality is the pervasive use of panoramic photography. Panoramic photography refers to a technique of presenting pictorial information with a prolonged field of vision (Jacobs, 2004). Using specialized equipment and computer programs, 360° panoramic pictures can entirely capture and portray the spatial information of a particular landscape (Jacobs, 2004). One 360° panoramic picture fully rotating a visual field contains the same amount of information as several still pictures capturing different angles of a landscape. By extending tourists' viewing angle and connecting each segment of pictorial information, 360° panoramic pictures provide an approximate reality of looking around a landscape. Hence, for online travel marketing, a higher level of modality, such as seamlessly integrated visual images, video clips, and 360° panoramic pictures, becomes a cost-efficient substitute for real travel experience (Cho et al., 2002; Tussyadiah & Fesenmaier, 2009). Although studies have supported that pictorial information in travel media substantially influences tourists' destination image formation (Cormany & Baloglu, 2011; MacKay & Fesenmaier, 1997), it is not yet clear how the mode of presenting pictorial information is related to the virtual travel experience. To compare the effects of different presentation modes, modality in this study is operationalized as the type of photography used to present the visual information of a tourist destination: still pictures and 360° panoramic pictures.

Researchers in media effect studies postulate that the psychological effects of technology affordance should be considered in multiple aspects because a single technical feature can produce both positive and negative user experiences (Balakrishnan & Sundar, 2011; Reeves & Nass, 1996; Sundar, 2008). For instance, moving or animated images can be beneficial to increase the level of attention but may also cause a feeling of dizziness. From this perspective, the unique effects of modality on four dimensions of virtual travel experience are separately proposed in the following.

Modality is an important technology affordance that influences the quality of the virtual travel experience because it has the potential to induce the "being there" heuristic (Sundar, 2008, p. 81). Media richness theory posits that a richer communication mode can increase a sense of realism (Daft & Lengel, 1984). Huang, Backman, Backman, and Chang (2016) also propose that a higher level of modality (e.g. 3D virtual reality) is considered to produce a stronger immersive user experience than a lower level of modality (e.g. two-dimensional photography). In a similar way, panoramic pictures providing richer spatial information with the connected visuals of landscapes may evoke the "being there" heuristic, which possibly leads to a higher sense of being located in the mediated environment than is the case with still pictures. Thus, the effect of modality on spatial self-location is hypothesized as follows:

H1a: Panoramic pictures of a tourist destination will lead to a greater spatial self-location than still pictures.

Novelty and exploration have been regarded as an important motivation factor fulfilling pleasure in a travel experience (Bello & Etzel, 1985). Presenting vivid and hyper-realistic visual images, newly developed modality technologies influence individuals' affection (Sundar, Narayan, Obregon, & Uppal, 1998). Related to users' affective response to advanced media technology, Sundar (2008) proposes that a higher level of modality can increase a positive feeling and immersion by producing "coolness" or "novelty" heuristics. A new type of presentation mode providing richer travel information (e.g. augmented reality) becomes a useful tool to elicit a positive attitude (Chung, Lee, Kim, & Koo, 2017). In a sense that panoramic photography is a relatively new and still evolving technical feature applied to online travel media, panoramic pictures may provoke a more immersive experience than still pictures.

H1b: Panoramic pictures of a tourist destination will lead to a greater affective engagement than still pictures.

When processing pictorial information, people focus on visual images on screen. When the visual images are moving, people's eyes need to follow them. In this case, a moving focal lens may produce the negative feelings of dizziness (Zhang et al., 2013). In their experiments, Kuze and Ukai (2008) found that users, especially less experienced one, feel dizziness and nausea when tracking moving images or following a focal lens. Similarly, Sundar (2008, p. 81) argues that a higher level of modality has the potential to elicit the "distraction" heuristic. In the real world, humans feel dizziness or nausea when their eyes track fast moving subjects. By applying the laws of physics to a media environment, the media equation theory suggests that looking at moving images on screen can also cause a feeling of dizziness (Reeves & Nass, 1996). Locating moving objects in a 3D space, stereoscopic 3D-TV viewing would produce the negative feelings of discomfort and sickness compared to traditional 2D TV viewing (Brunnström, Wang, Tavakoli, & Andrén, 2017). Because 360° panoramic pictures are created by a full rotation of a vision field, it is necessary to follow a moving focal lens to see the entire image of panoramic pictures. Compared to the static viewpoint of still pictures, the moving images of a landscape in panoramic pictures may cause a higher level of distraction or dizziness.

H1c: Panoramic pictures of a tourist destination will lead to greater disorientation than still pictures.

For the cognitive dimension of virtual travel experience, modality is evidently related to the formation of a spatial mental model (Reeves & Nass, 1996; Wirth et al., 2007). Researchers in media effect studies postulate that motion on the screen has the power to change the cognitive dimension of a virtual experience (Balakrishnan & Sundar, 2011; Reeves & Nass, 1996). As motion is an essential characteristic of the real world, Reeves and Nass (1996, p. 219) argue that motion on the screen increases users' attention and memory by inducing "a visual orienting response". In addition, the extension of the field of vision helps users to better understand the spatial structure of a virtual environment (Balakrishnan & Sundar, 2011). Similarly, media richness theory posits that a higher level of presentation mode may establish a better setting for approximate reality to provide accurate communication information (Daft & Lengel, 1984). Because of the enhanced cognitive attention on moving images and the seamless connection of pictorial information, users of panoramic pictures can more easily understand and interpret the spatial structure of a virtual environment than users of still pictures.

H1d: Panoramic pictures of a tourist destination will lead to a greater spatial mental model than still pictures.

Impact of navigability on virtual travel experience

Navigability is an affordance that determines the degree of users' movement in a web interface (Sundar, 2008). Because online travel media construct navigable web structures with hyper-linked webpages and multilevel layers, tourists can browse and explore a virtual environment. From the perspective of human–computer interaction, Balakrishnan and Sundar (2011) conceptualize navigability based on the 2D concept of navigation: travel and way-finding. Travel refers to physical movement from one location to another location, and way-finding represents the cognitive process of finding the way to a designated location. Corresponding to the dimensionality of navigation, Balakrishnan and Sundar (2011) articulate two aspects of navigability: traversibility and guidance. Reflecting the dimension of travel, traversibility is defined as the level of controlling a user's movement and viewpoint in a virtual environment. In online travel media technical features such as arrows to move forward or backward, clicking to transport into a different landscape, or mouse dragging to spin a user's viewpoint, are associated with traversibility. In terms of way-finding, guidance indicates the level of scaffolding, which helps tourists understand the structure of a virtual environment and to find directions. Technical features of guidance in online travel media include map functions or indicators of a user's current and designated location.

Researchers have proposed that recognizing or using navigability can influence media users' perceived quality, attitude, and behavior (Balakrishnan & Sundar, 2011; Sundar, 2008; Sundar et al., 2012). The mere presence of visual cues of navigability (e.g. navigation buttons and arrows) is capable of generating certain types of heuristics (e.g. browsing, elaboration, scaffolding, and prominence), which influence the perceived quality of information and the virtual experience (Sundar, 2008). In addition to the visual cue effect, researchers have examined actual use effect in a sense that media users can experience the full potential of navigability features by using them (Sundar et al., 2012; Wu, 2010). When exploring navigable web interfaces, media users tend to consider information search for tangible retail products (e.g. computers and clothing) more enjoyable and immersive than non-navigable web interfaces (Fiore, Kim, & Lee, 2005; Suh & Chang, 2006; Wu & Chang, 2005). Similarly, for online travel media recognizing and/or actually using navigability features would enrich the virtual travel experience. Hence, navigability in this study is operationalized as the existence of technological features that provide the possibility of controlling users' movement and providing guidance functions in a virtual environment. The impacts of navigability on dimensions of virtual travel experience are proposed in the following.

Empirical studies have supported a positive impact of navigability on virtual travel experience (Balakrishnan & Sundar, 2011; Hyun & O'Keefe, 2012). Particularly for spatial self-location, navigability may have the power to produce the feeling of being there (Balakrishnan & Sundar, 2011; Hyun & O'Keefe, 2012; Sundar, 2008). Navigability features would positively influence tourists' perception of movement when recognizing the possibility of movement or actually moving around in a virtual environment. Because navigability can increase the sense of exploring and moving around, tourists may develop a feeling of being located in a virtual environment. Also, when using navigability features, tourists can experience a sense of control. Studies on video gaming and virtual reality have supported the argument that a sense of control is a central element in producing a feeling of being in a virtual environment (Klein, 2003; McMillan & Hwang, 2002). Hence, presence of navigability is expected to enhance tourists' spatial self-location by allowing them to control their movement.

H2a: Presence of navigability will lead to a greater spatial self-location than absence of navigability.

The presence of navigability provides users with a broader scope of choices and control over their movement in a virtual environment (Balakrishnan & Sundar, 2011; Sundar, 2008). According to the enhanced affective engagement model, the existence of choice positively influences media users' affective engagement in completing a required task (Cordova & Lepper, 1996; Flowerday & Schraw, 2003; Schraw, Flowerday, & Reisetter, 1998). Similarly, media flow theory posits that users' level of control and the corresponding immediate feedback can lead to a flow-like experience of being fully immersed in a given activity (Sherry, 2004). Empirical evidence supports that user control increases an individual's emotional engagement when using media (Flowerday & Schraw, 2003; Schraw et al., 1998). For online travel media, navigability inherently involves a certain level of freedom to control users' movement in a virtual environment. Thus, presence of navigability is expected to enhance tourists' affective engagement in using online travel media.

H2b: Presence of navigability will lead to greater affective engagement than absence of navigability.

Navigability is related to users' power to manage the volume of information process by changing the steering angle of visual images or controlling a moving focal lens on screen. For instance, the presence of navigability allows users to control the degree of changing a visual field and the speed of looking around a landscape. Using navigability features, tourists can decide on the speed with which a focal lens moves, fast or slow, as they explore in a virtual environment. Synthesizing recent studies on video gaming, Crick (2010) argued that a higher level of control in first-person visual mode reduces the feeling of dizziness while increasing the sense of presence and affective engagement. Thus, this study hypothesizes that navigability may reduce tourists' feeling of disorientation.

H2c: Presence of navigability will lead to lesser disorientation than absence of navigability.

Navigability is also related to the cognitive dimension of virtual travel experience. Related to spatial mental modeling, navigability provides the flexibility of processing the amount of pictorial information. Using navigability features, tourists can better understand the spatial structure of a destination by spending more time on things they are interested in (Balakrishnan & Sundar, 2011). In addition, because navigability is the action-oriented technical feature that requires a certain level of attention to control, tourists may pay more attention to their movement when exploring a virtual environment (Reeves & Nass, 1996). According to the enhanced cognitive engagement model, the existence of choice is positively associated with an individual's attention to a given learning activity (Flowerday & Schraw, 2003). The increased attention will positively influence the level of understanding the spatiality of a virtual environment. By providing the possibilities of focusing on particular pictorial information, navigability may increase tourists' spatial mental model.

H2d: Presence of navigability will lead to a greater spatial mental model than absence of navigability.

Relation between virtual travel experience and behavioral intention

Researchers have suggested that there is a positive relation between virtual travel experience and tourists' behavioral intention to visit and recommend to others (Hyun & O'Keefe, 2012; Kang & Gretzel, 2012). Affect transfer models suggest that particular emotions and feelings generated from the mediated experience are transferred to an individual's attitude toward a product or service (Homer, 2006; Machleit, Allen, & Madden, 1993). Similarly, in tourism settings, there is empirical evidence supporting a positive relation between the emotional perception of touristic experience and behavioral intention (Lee, Lee, & Choi, 2011; Lee, Yoon, & Lee, 2007). In their study on the communication processing models of online travel information, Tang and Jang (2012) assert that affective dimensions of experiencing a travel website are positively associated with tourists' attitude toward a destination and behavioral intention to visit. In this regard, this study proposes that the affective dimensions of virtual travel experience (i.e. spatial self-location, affective engagement, and disorientation) may have substantial effects on behavioral intention to visit and to recommend to others.

H3a: Greater spatial self-location will lead to greater behavioral intention.

H3b: Greater affective engagement will lead to greater behavioral intention.

H3c: Greater disorientation will lead to lesser behavioral intention.

Tourism researchers have suggested that the cognitive process of evaluating a tourism product or service is a primary determinant for satisfaction and behavioral intention (Lee et al., 2011, 2007). When exploring a travel website, cognitive elaboration of using high-quality information has a positive impact on attitude and behavioral intention (Tang & Jang, 2012). Hence, this study assumes that comprehensive understanding of a destination (spatial mental model) is positively related to behavioral intention. Figure 1 illustrates the proposed relationships of

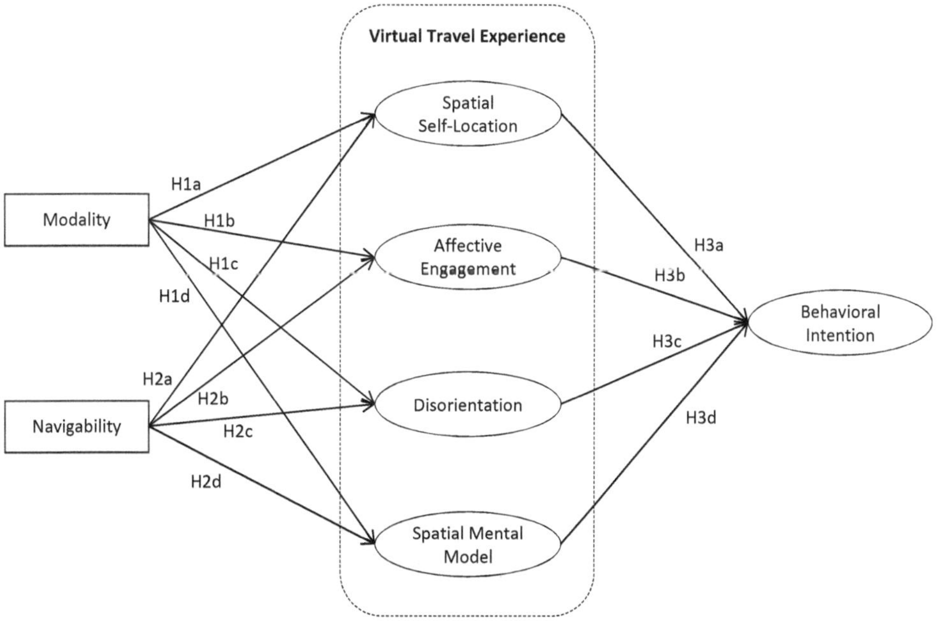

Figure 1. Proposed relationships of technology affordances, virtual travel experience, and behavioral intention.
H: hypothesis.

technology affordances, four dimensions of virtual travel experience, and behavioral intention.

H3d: Greater spatial mental model will lead to a greater behavioral intention.

Method

Research design

In order to examine the proposed relationships of technology affordances, virtual travel experience, and behavioral intention, a 2 (modality: still pictures versus panoramic pictures) × 2 (navigability: absence versus presence) between-subjects experiment was designed. This study took place in a computer-laboratory at a large university in the United States (US).

Participants

Initially, 218 participants were recruited from undergraduate classes in exchange for extra course credit. After excluding five cases due to non-adherence to the instructions or a large portion of missing data, 213 individuals aged 18–26 years (M = 20.28, SD = 1.23) were used for analysis. There were more female (N = 157, 73.71%) than male participants (N = 56, 26.29%).

Stimuli

In order to create the stimulus materials, a virtual travel web application (from www.airpano.com) portraying the visual landscapes of Easter Island, Chile, was used (Figure 2). This web application contains a map of the island and 13 360° panoramic pictures taken of actual sites (e.g. Moai statues, quarries, and other natural environment). With built-in navigability features, tourists can navigate and explore different touristic sites by either mouse-dragging or clicking navigation buttons on the map and pictures.

Manipulation

Modality was manipulated by converting 13 panoramic pictures into 52 still pictures (13 × 4 = 52) to maintain the same amount of visual information of the destination. For instance, participants in the still pictures condition watched 52 still pictures whereas participants in the panoramic pictures condition watched 13 panoramic pictures. Navigability was manipulated by disabling (absence of navigability) or enabling navigability features (presence of navigability). Participants in the absence of the navigability condition passively watched automatically rotating pictures, while participants in the presence of the navigability condition were allowed to use a mouse to click or drag to move around or see different angles of the sites.

In order to examine the effects of modality and navigability, different web interfaces were created for four conditions (condition 1 = still pictures and absence of navigability; condition 2 = still pictures and presence of navigability; condition 3 = panoramic pictures and absence of navigability; condition 4 = panoramic pictures and presence of navigability). All participants were instructed to use a demonstration of a new travel website to explore the island. They were asked to explore all 13 places marked by Moai icons (Table 1). In condition 1, participants watched still pictures that were automatically changed every 5 seconds. Participants in condition 2 were able to click icons they wanted to see in order, and were then told to use arrows to see different angles of each site with four still pictures. For condition 3, participants watched 13 panoramic pictures that automatically changed and rotated for 20 seconds. Participants in condition 4 were allowed to click icons they wanted to see in order, and were then able to click or mouse-drag to see different angles of each site with a panoramic picture. After creating the stimuli, a pilot test with five participants for each condition was conducted to identify and resolve potential issues with regard to the stimuli.

Measures

A post-experimental questionnaire was used to measure participants' touristic experience in the virtual environment, including spatial self-location, affective engagement, disorientation, and spatial mental model and behavioral intention to visit and recommend. For spatial self-location, three items (e.g. "I felt as though I was physically present in this place") were adopted from the Spatial Presence Questionnaire (Vorderer et al., 2003). Derived and modified from Flowerday and Schraw (2003), 11 items were used to measure affective engagement (e.g. "I enjoyed participating in this study"). Operationalized as the degree of dizziness felt during the exploration, disorientation was measured with four items (e.g. "This web interface made me feel dizzy") from Zhang et al. (2013). Spatial mental model was measured with five items (e.g. "I was able to imagine the arrangement of the sites in this place very well") derived from the Spatial Presence Questionnaire (Vorderer et al., 2003). In terms of behavioral intention (e.g. "I would visit the tourist destination I viewed on the website"), four items were adopted from the work

Figure 2. Snapshots of the stimuli (map of Easter Island and a still picture of a location).

of Zeithaml, Berry, and Parasuraman (1996). All items for spatial self-location, affective engagement, disorientation, spatial mental model, and behavioral intention were measured on a 7-point Likert scale (1 = strongly disagree, 7 = strongly agree). This study also measured prior knowledge of a destination as a control variable that may influence behavioral intention by using three items (previous experience, familiarity, and expertise) derived from Kerstetter and Cho (2004). A full list of the items from each measure can be viewed in Table 1.

Procedure

Individuals were invited to participate in this study by signing up for one of 50 time slots from May to August 2013. Randomly assigned to one of the four experimental conditions, each time slot was independently operated with 2–10 participants. Upon their arrival at the computer laboratory, participants were asked to fill out a consent form and pre-experimental questionnaire with paper and pencil recording their demographic information. Then they were briefed about the overall experimental procedure and instructed that they would be either watching or exploring a demonstration of a new travel website. Then, basic information about the destination was given: "You will explore Easter Island in Chile. Easter Island is famous for its 887 monumental statues, called Moai. This island is a UNESCO world heritage site." The instruction to use each web interface was delivered to participants by the same researcher. Then, participants experienced the assigned web interface using one of the desktop personal computers in the laboratory (full screen on a 17-inch

Table 1. Results of confirmatory factor analysis for the virtual travel experience dimensions and behavioral intention.

Items	FL[a]	t	SE
Spatial Self-Location			
I felt like I was actually there on the island.	.91	NA[b]	NA
I felt as though I was physically present on the island.	.86	16.26	.06
It seemed as though I actually took part in action on the island.	.81	14.84	.06
Affective Engagement			
I appreciate the choices I got to make in this study.	.85	NA	NA
I liked what I was asked to do in this study.	.74	12.00	.08
I was highly motivated to participate in this study.	.71	11.49	.08
I enjoyed participating in this study.	.69	11.06	.07
I could easily identify with what I see.	.64	9.96	.09
I felt I had a great deal of control in this study.	.61	9.29	.09
I tried really hard in this study.	.44	6.47	.10
Disorientation			
I felt I had a headache.	.89	17.48	.05
I felt dizzy.	.88	NA	NA
I felt nauseous.	.81	15.00	.04
I had eyestrain.	.75	13.23	.06
Spatial Mental Model			
I was able to imagine the arrangement of the sites on the island very well.	.78	10.00	.09
I had a precise idea of the spatial surroundings on the island.	.74	9.64	.09
I was able to make a good estimate of how far apart things were from each other on the island.	.72	NA	NA
I was able to make a good estimate of the sizes of objects on the island.	.66	8.63	.10
I still have a concrete mental image of the island.	.57	7.48	.08
Behavioral Intention			
I would recommend the tourist destination I viewed on the website to a friend.	.87	NA	NA
I would say positive things about the tourist destination.	.84	15.05	.05
I would visit the tourist destination I viewed on the website.	.82	14.64	.06
I would seek more information about the tourist destination I viewed on the website.	.73	12.37	.06

Note. [a]FL = factor loading, t = t-value, SE = standard error; [b]t-values for parameters fixed at 1.0 for identification purposes were not available (NA)

monitor). Participants in conditions 1 and 3 (M = 390.00 seconds, SD = .00) spent more time exploring the given web interface than participants in condition 2 (M = 347.50 seconds, SD = 75.73) and 4 (M = 366.48 seconds, SD = 57.01). After they finished exploring the web interface, participants were asked to fill out a post-experimental online questionnaire using Qualtrics Online Survey software.

Results

The data were screened for normality and missing values. Based on the shape of the histogram and the range of kurtosis (|<|10) and skewness (|<|3), the variables followed a normal distribution (Kline, 2011). Less than 5% of the data was missing (.0–.9%); thus, missing values were not considered a problem and substituted with an expectation-maximization procedure (Tabachnick & Fidell, 2000).

Spatial self-location, affective engagement, disorientation, spatial mental model, and behavioral intention were measured in this study. A confirmatory factor analysis was conducted to examine reliability and validity of the five dimensions. Based on model fit indices (χ^2 = 520.064, df = 220, p = .000; RMSEA = .082, 90% CI = .075-.089; NNFI = .91; CFI = .90), the initial measurement model had a marginally acceptable model fit (Hu & Bentler, 1999). However, after four items in affective engagement were eliminated iteratively because the modification indices showed that one item had an unacceptably weak factor loading less than .4 (Kline, 2011) and three items had high correlations of errors (Kline, 2011), the model fit of the final measurement model was substantially improved (χ^2 = 404.064, df = 220, p = .000; RMSEA = .063, 90% CI = .056-.070; NNFI = .92; CFI = .93). All factor loadings (λ = .57–.91) were statistically significant (see Table 1).

As shown in Table 2, each construct had a high composite reliability (CR) (.83–.90) and acceptable average variance extracted (AVE) (.44–.73). Discriminant validity was also confirmed as each measure had a higher value of AVE than the square of the inter-correlation between corresponding measures (Fornell & Larcker, 1981).

In order to examine the effects of modality and navigability on the dimensions of virtual travel experience, a series of 2 (modality) × 2 (navigability) analyses of covariance (ANCOVAs) was conducted (Table 3). Regarding the potential effects on virtual travel experience, variances for age, gender, exploration time, and prior knowledge were statistically controlled. Among the three items measuring prior knowledge, only two items capturing familiarity and expertise were used to create a construct because there was no participant who had actually visited Easter Island. Because familiarity/expertise was substantially positively skewed, a new construct (x`) was created; x` = log10(x) (Tabachnick & Fidell, 2000).

In terms of spatial self-location, the results revealed a reliable main effect for modality (F = 7.73, df = 1, p < .01). Spatial self-location was significantly higher in the condition of panoramic pictures (M = 4.79, SD = 1.35) compared to the condition of still pictures (M = 4.30, SD = 1.25). Thus,

Table 2. Correlations, composite reliability, AVE, Mean, and SD.

	SSL	AE	DIS	SMM	BI
SSL	1.00				
AE	.44	1.00			
DIS	−.17	−.39	1.00		
SMM	.48	.52	−.31	1.00	
BI	.52	.60	−.27	.43	1.00
CR	.89	.84	.83	.90	.89
AVE	.73	.44	.70	.49	.67
Mean	4.54	5.56	2.23	5.07	5.60
SD	1.32	.86	1.36	1.04	1.13

SSL: spatial self-location; AE: affective engagement; DIS: disorientation; SMM: spatial mental model; BI: behavioral intention; CR: composite reliability; AVE: average variance extracted; SD: standard deviation.

Table 3. Results of two-way ANCOVAs on virtual travel experience.

	Sum of Squares	df	F	Sig.
Spatial self-location				
Modality	12.55	1	7.73	.00**
Navigability	5.48	1	3.38	.07
Modality * Navigability	8.23	1	5.07	.03*
Age	1.84	1	1.13	.29
Gender	2.40	1	1.48	.23
Exploration time	1.35	1	.83	.36
Prior knowledge	.70	1	.43	.51
Affective engagement				
Modality	2.08	1	2.97	.09
Navigability	2.83	1	4.03	.05*
Modality * Navigability	3.32	1	4.73	.03*
Age	.20	1	.29	.59
Gender	.00	1	.00	.95
Exploration time	.41	1	.58	.45
Prior knowledge	2.63	1	3.74	.05
Disorientation				
Modality	70.12	1	48.15	.00**
Navigability	7.09	1	4.87	.03*
Modality * Navigability	1.69	1	1.16	.28
Age	2.74	1	1.88	.17
Gender	17.83	1	12.24	.00*
Exploration time	.00	1	.00	.96
Prior knowledge	.89	1	.61	.44
Spatial mental model				
Modality	4.02	1	3.91	.05*
Navigability	10.18	1	9.89	.00**
Modality * Navigability	.00	1	.00	.97
Age	.46	1	.45	.50
Gender	1.82	1	1.77	.19
Exploration time	.00	1	.00	.94
Prior knowledge	3.53	1	3.43	.07

*$p < .05$; **$p < .01$; ANCOVA: analysis of covariance.

H1a was supported. The main effect for navigability was marginally significant (F = 3.38, df = 1, p = .07). Interestingly, the interaction on spatial self-location was significant (F = 5.07, df = 1, $p < .05$). As depicted in Figure 3, when panoramic pictures were presented, spatial self-location was higher in the presence of the navigability condition (M = 5.16, SD = 1.23) than in the absence of the navigability condition (M = 4.39, SD = 1.37). However, when still pictures were presented, there was no significant difference between participants in the absence or presence of the navigability condition, showing that H2a was partially supported. Control variables were not significant.

The results for affective engagement revealed that the main effect for modality (F = 2.97, p = .09) was marginally significant (H1b was not supported), and the main effect for navigability (F = 4.04, df = 1, $p < .05$) was significant, supporting H2b. Participants in the presence of the navigability condition (M = 5.71, SD = .84) reported a higher affective engagement than those in the absence of the navigability condition (M = 5.41, SD = .86). Similar to spatial self-location, the interaction on affective engagement was significant (F = 4.73, df = 1, $p < .05$). As shown in Figure 4, when panoramic pictures were presented, affective engagement was higher in the presence of the navigability condition (M = 5.73, SD = .73) than in the absence of the navigability condition (M = 5.19, SD = .87). However, when still pictures were presented, there was no significant difference between participants in the absence or presence of the navigability condition. Control variables were not significant.

With regard to disorientation, the results revealed that the main effect for modality (F = 48.15, df = 1, $p < .01$) as well as the main effect for navigability (F = 4.87, df = 1, $p < .05$) were significant. Both H1c and H2c were supported. Participants viewing panoramic pictures (M = 2.76, SD = 1.52) felt more disorientation than those viewing still pictures (M = 1.71, SD = .94). However, the presence of navigability (M = 1.99, SD = 1.13) significantly reduced participants' feeling of disorientation compared to the absence of navigability (M = 2.49, SD = 1.56). The interaction effect was not significant. Among control variables, only gender had a significant effect (F = 12.24, df = 1, $p < .01$). Female participants (M = 2.37, SD = 1.41) reported a higher level of disorientation than males (M = 1.84, SD = 1.14).

In terms of the spatial mental model, the main effects for both modality (F = 4.02, df = 1, $p < .05$) and navigability (F = 10.18, df = 1, p < .01) were

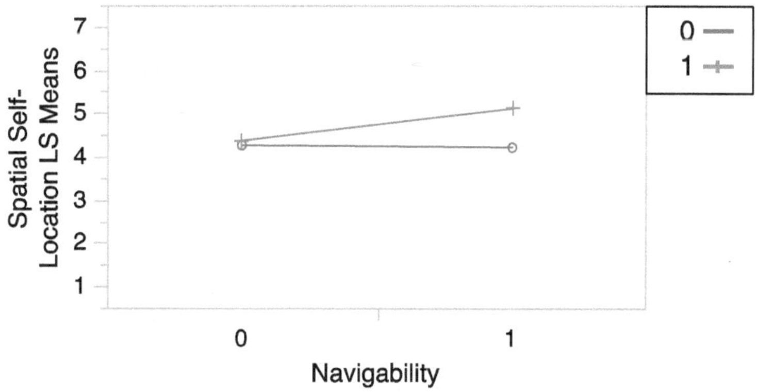

Figure 3. Interaction effect of modality and navigability on spatial self-location.
0 = still pictures; 1 = panoramic pictures.

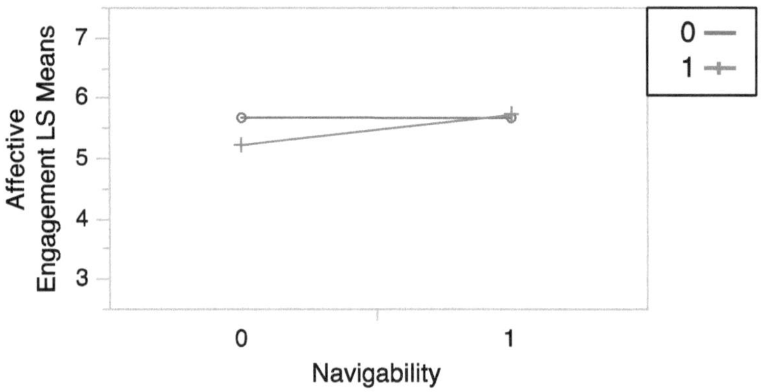

Figure 4. Interaction effect of modality and navigability on affective engagement.
0 = still pictures; 1 = panoramic pictures.

significant, supporting H1d and H2d. As predicted, the spatial mental model was higher in panoramic pictures (M = 5.19, SD = 1.00) than in still pictures (M = 4.95, SD = 1.07). Also, participants reported a higher spatial mental model in the presence of navigability (M = 5.29, SD = .96) compared to the absence of navigability (M = 4.84, SD = 1.08). However, the interaction and control variables were not significant.

In order to examine the relationship between virtual travel experience and behavioral intention, a hierarchical multiple regression analysis was conducted. Control variables (age, gender, major, exploration time, and familiarity/expertise) were included in the first-step model. As shown in Table 4, only familiarity/expertise had a significant positive effect and the effects of other variables were not significant. The five control variables accounted for 5% of the variance in behavioral intention. The second step of the model included the four constructs of the virtual travel experience. Spatial self-location and affective engagement had a significant effect whereas the effects of disorientation and the spatial mental model were not significant. Thus, H3a and H3b were supported. H3c and H3d were not supported. The final model explained 40% of the variance in behavioral intention, an increase of 35% with the inclusion of the four virtual travel experience dimensions.

Table 4. Hierarchical multiple regression model on behavioral intention.

Independent variable	B	S.E.	*t*-value
Step 1			
Age	.05	.72	.90
Gender	−.10	1.19	1.49
Major	0.04	.18	.23
Exploration time	.00	.11	.14
Familiarity/Expertise	.81	6.43	8.04**
R^2 = .05			
Step 2			
Spatial self-location	.26	18.59	23.22**
Affective engagement	.47	22.12	27.63**
Disorientation	−.02	.14	.18
Spatial mental model	.05	.46	.58
R^2 = .40; R^2 Change = .35			

*p < .05; **p < .01; LS: least squares

Further, considering the sequential relationships between technology affordance, virtual travel experience, and behavioral intention, the mediating role of virtual travel experience was examined. Based on the interaction effects of modality and navigability on spatial self-location and affective engagement, the mediation effects for these two constructs were tested using moderated mediation models (Hayes, 2013). For spatial self-location, when still pictures were presented, the mediation effect was not significant (indirect effect = .02; CI = -.19 – .20). However, when panoramic pictures were presented, spatial self-location mediated the effect of navigability on behavioral intention (indirect effect = .32; CI = .11 – .55). Similarly, the mediation effect of affective engagement on the relationship between navigability and behavioral intention was significant with panoramic pictures (indirect effect = .37; CI = .16 – .62), but not still pictures (indirect effect = .04; CI = -.19 – .25). The mediation roles of disorientation and spatial mental model were not supported. Given that two constructs of virtual travel experience had a significant mediation effect for the panoramic pictures condition, a multiple mediation model was applied to compare the mediation effects of spatial self-location and affective engagement. By selecting only participants who received panoramic pictures, the mediating roles of spatial self-location and affective engagement on the relationship between navigability and behavioral intention were examined with bootstrapping procedures using 2000 bootstrap samples and bias-corrected confidence intervals. This analysis revealed significant indirect effects for both mediators. The mediation effect for affective engagement was stronger than that for spatial self-location. The results of this study are summarized in Figure 5.

Discussion

For online travel media, technology affordances determine how tourists experience the pictorial information of a destination (Balakrishnan & Sundar, 2011; Huang et al., 2016). Focusing on the mode of presenting pictorial information and the possibility of movement in a virtual environment, the purpose of this experiment was to understand the effects of modality and navigability on virtual travel experience and behavioral intention. The distinctive psychological effects of modality and navigability were examined for both affective (i.e. spatial self-location, affective engagement, and disorientation) and cognitive (i.e. spatial mental model) dimensions of virtual travel experience. Furthermore, by scrutinizing the mediating role of virtual travel experience, this study attempted to explain the psychological mechanism of how modality and navigability influence tourists' behavioral intention to visit and recommend to others.

By supporting the effects of affordances on touristic experience, the findings provide theoretical contributions to the tourism literature. First, this study provides empirical evidence that not only the quality or content of pictorial information, but also technical features of online travel media, can elicit substantial psychological effects. As McLuhan (1964) argues, "the medium is the message", this study proposes that the technical form of online travel media influences how tourists perceive the pictorial information of a destination. Consistent with previous research on the technical form of web interfaces and user experience (Hyun & O'Keefe, 2012; Wu & Chang, 2005), the results indicated that several dimensions of virtual travel experience were influenced by affordances in constructing the spatial structure of a destination in a virtual environment. Also, this study adopted the variable-centered approach (Reeves & Nass, 1996; Sundar, 2009) to differentiate the effects produced by modality and navigability. By supporting that each affordance conveys different types of heuristics (Sundar, 2008), the results revealed that the psychological effects of modality and navigability are distinctive depending on the dimensions of virtual travel experience.

For spatial self-location, panoramic pictures provided an enhancement of tourists' feeling of being located in a virtual environment compared to still pictures. In a sense that a richer presentation mode is associated with a higher level of realism, this result is consistent with media richness theory (Daft & Lengel, 1984). More importantly, this study found that the effect of navigability is dependent upon the condition of modality. For instance, the presence of navigability features increased spatial self-location only when panoramic pictures were presented. This result implies that navigability is useful to elicit the "being there" effect for a higher level of presentation mode that constructs a seemingly connected virtual environment. Similarly, this study revealed the interaction effect on affective engagement. With still pictures, affective engagement was high regardless of the existence of navigability features. However, when using panoramic pictures, affective engagement was significantly decreased in the absence of the navigability condition. Because they are unlike still pictures, 360° panoramic pictures inherently involve a moving focal lens and rotating image of a landscape; thus, participants may expect that panoramic pictures contain navigability features. This result suggests that navigability is considered to be a necessary condition

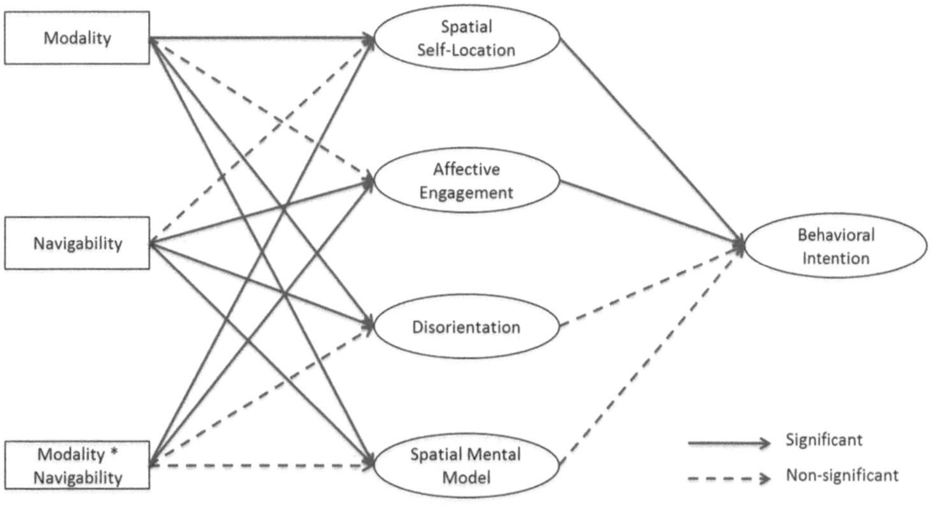

Figure 5. Results of the relationships between modality, navigability, virtual travel experience, and behavioral intention.

for panoramic pictures to elicit participants' affective engagement in a given task.

Given that panoramic pictures produced a greater feeling of disorientation than still pictures, it is important to note that a higher level of modality is not always beneficial to enhancing the quality of virtual travel experience. This finding indicates that panoramic pictures seem to elicit distraction heuristics (Sundar, 2008). However, navigability was useful in reducing the level of disorientation, suggesting that user control is a fundamental feature to reconcile the negative affection of dizziness and distraction for the computer-mediated environment. Interestingly, there was a significant gender difference: female participants felt a higher level of disorientation than male participants. This may be because males tend to have more experience using panoramic images and 3D simulations embedded in video games and other computer tasks (Feng, Spence, & Pratt, 2007; Okagaki & Frensch, 1994).

For the cognitive dimension of the virtual travel experience, this study found that both modality and navigability have significant effects on spatial mental models. More specifically, panoramic pictures were beneficial to better understanding the structure of the virtual environment compared to still pictures. This result supports the argument that motion on the screen (Reeves & Nass, 1996) and the connection of pictorial information (Balakrishnan & Sundar, 2011) in panoramic pictures are helpful in conveying more accurate information about the landscapes of a destination. The presence of navigability features also led to a greater spatial mental model. Providing the possibility of movement, navigability often contains action-oriented technical features (Sundar, 2008). As the enhanced cognitive engagement model argues (Flowerday & Schraw, 2003) and the results of this study suggest, the availability of choice allowed by navigability can increase individuals' attention in a given task and their cognitive ability.

With regard to the current scholarly discourse on the relationships between virtual, imaginative, and corporeal travel (Reijnders, 2011; Urry, 2000), this study's findings provide preliminary evidence that the positive affect of virtual travel experience can be transferred to corporeal travel. More specifically, spatial self-location and affective engagement had significant positive effects on behavioral intention. Many tourism studies have shown that positive affective responses (e.g. emotional value, enjoyment, and satisfaction) elicited by a tourism service/product are strongly associated with tourists' behavioral intention and future behavior (Lee et al., 2011, 2007). Similarly, in the online context, Tang and Jang (2012) argue that the affective dimensions of experiencing travel websites can be easily transferred to an individual's attitude toward a website or destination, and travel intention. Uniquely, this study revealed the moderated mediation effects of modality and navigability on spatial self-location and affective engagement. Only when panoramic pictures were presented, positive affects (spatial self-location and affective engagement) mediated the effect of navigability on behavioral intention. Emphasizing the mediating role of positive affects for a higher modality of presenting pictorial information, this result helps to explain the underlying mechanism of how tourists' behavioral intention is influenced by navigability features.

However, the negative affect and cognitive dimension of virtual travel experience had no significant effects on behavioral intention in this study. In terms of disorientation, although there was a significant difference, the mean scores (M = 1.58–3.14; 7-point scale) in all four experimental conditions were low. In this regard, it is possible that participants might have felt a negligible level of disorientation to influence their behavioral intention. The affective–cognitive model of consumer decision-making (Shiv & Fedorikhin, 1999) posits that an individual's behavior is primarily influenced by the affective responses elicited by a given task if the availability of information is limited. Because participants only spent about 6.5 minutes on average on the given task, which mainly conveyed pictorial information, their behavioral intention could have been strongly influenced by affective responses, not the cognitive processing of recognizing and understanding symbols and spatial cues of a destination. In addition to the effects of virtual travel experience, this study supports previous research by Kerstetter and Cho (2004) that tourists' previous experience with a destination is strongly related to their behavioral intention. Yet, the findings are limited in terms of articulating a threshold when significant effects would occur because there were only two levels of modality and navigability. The student sample may enhance the internal validity of the experiment, but it also weakens the external validity.

This study provides several useful practical implications. First, tourism marketers and DMOs should keep in mind that the technical form of online travel media can induce substantial effects on mediated travel experience and behavioral intention. Modality and navigability are the main technology affordances accelerating visual innovation trends in online travel media. This study shows that modality and navigability shape and enhance the quality of mediated travel experience, providing ways of gazing upon and interacting with spatial and cultural landscapes of a destination. As online travel marketing becomes more competitive, adopting innovative visual technologies (Horng & Tsai, 2010;

Lepp et al., 2011; Michaelidou et al., 2013), it is important to understand the pivotal role of modality and navigability in producing a positive touristic experience and increasing behavioral intention. In a sense that users' positive affection (spatial self-location and affective engagement) is significantly related to behavioral intention, newly developed modality, such as 3D virtual reality (VR) or augmented reality (AR), would be beneficial for online travel media to present a destination's visual images. Especially for the millennial generation and technology-savvy tourists, the mediated experience through the lens of AR/VR can be powerful enough to change their perceptions of the destination.

Second, the role of modality and navigability is applicable to the design of visual interfaces. Dissecting and focusing on fundamental technical components, the findings are not limited to the available technologies, but are also applicable to the design of newly emerging technologies, such as VR/AR simulations or near-field communication applications. For website designers or mobile application developers, this study's results suggest that adding new technology attributes is not necessarily beneficial. Instead, the psychological effects of each technical attribute should be considered in terms of multiple dimensions of mediated experience (Reeves & Nass, 1996). Web designers should carefully consider how tourists emotionally and cognitively respond to the different modes of visual images and navigability features when exploring a virtual environment.

Last but not least, it is important to understand and predict the combined effects of two or more technology affordances (Sundar, 2008). Considering the negative feeling of disorientation, a higher level of modality with moving images and a focal lens should be developed with built-in navigability features. For many VR simulations and devices it becomes an important technical agenda to resolve the side effects of nausea, dizziness, and distraction (Freeman, 2016). This study provides a design guideline and empirical evidence showing how navigability features can be utilized to reduce the feeling of disorientation. As more people travel with personal mobile devices equipped with wireless connectivity and the global positioning system (GPS), the findings emphasize the important role of navigability for location-based AR simulations (Chung et al., 2015; Jung et al., 2015). Navigability will broaden the scope of the user experience, synthetically interacting with the physical environments of a destination.

Responding to the emergence of new media technology, future research should be conducted to incorporate more diverse and advanced modes to portray a destination. Building on Balakrishnan and Sundar's (2011) conceptualization of the two dimensions of navigability, future research should be designed to differentiate the effects of technical features associated with traversibility and guidance. In this study, the cognitive dimension of virtual travel experience was self-reported. Considering the functions of the spatial mental model, however, Balakrishnan and Sundar (2011) suggest that there is an evident difference between an individual's perception and his or her actual achievement of spatial information. For instance, individuals tend to be generous when evaluating their perceived understanding of spatial information, even if actual understanding of the given spatial information is poor. Thus, researchers in future studies should examine how modality and navigability relate to the perceived and achieved spatial mental model and how their influence on tourists' behavioral intention differs.

Disclosure statement

No potential conflict of interest was reported by the authors.

References

Balakrishnan, B., & Sundar, S. S. (2011). Where am I? How can I get there? Impact of navigability and narrative transportation on spatial presence. *Human Computer Interaction, 26*, 161–204.

Bello, D. C., & Etzel, M. J. (1985). The role of novelty in the pleasure travel experience. *Journal of Travel Research, 24*(1), 20–26. doi:10.1177/004728758502400104

Brunnström, K., Wang, K., Tavakoli, S., & Andrén, B. (2017). Symptoms analysis of 3D TV viewing based on Simulator Sickness Questionnaires. *Quality and User Experience, 2*(1), 1–15. doi:10.1007/s41233-016-0003-0

Cary, S. (2004). The tourist moment. *Annals of Tourism Research, 31*, 61–77. doi:10.1016/j.annals.2003.03.001

Cho, Y., Wang, Y., & Fesenmaier, D. R. (2002). Searching for experiences - The Web-based virtual tour in tourism marketing. *Journal of Travel & Tourism Marketing, 12*(4), 1–17. doi:10.1300/J073v12n04_01

Chung, N., Han, H., & Joun, Y. (2015). Tourists' intention to visit a destination: The role of augmented reality (AR) application for a heritage site. *Computers in Human Behavior, 50*, 588–599. doi:10.1016/j.chb.2015.02.068

Chung, N., Lee, H., Kim, J.-Y., & Koo, C. (2017). The role of augmented reality for experience-influenced environments: The case of cultural heritage tourism in Korea. *Journal of Travel Research*. doi:10.1177/0047287517708255

Cordova, D. I., & Lepper, M. R. (1996). Intrinsic motivation and the process of learning: Beneficial effects of contextualization, personalization, and choice. *Journal of Educational Psychology, 88*, 715–730. doi:10.1037/0022-0663.88.4.715

Cormany, D., & Baloglu, S. (2011). Medical travel facilitator websites: An exploratory study of web page contents and services offered to the prospective medical tourist. *Tourism Management, 32*, 709–716. doi:10.1016/j.tourman.2010.02.008

Crick, T. (2010). The game body: Toward a phenomenology of contemporary video gaming. *Games and Culture, 6*(3), 259–269. doi:10.1177/1555412010364980

Daft, R. L., & Lengel, R. H. (1984). Information richness: A new approach to managerial behavior and organization design. In B. M. Staw & L. L. Cummings (Eds.), *Research in organizational behavior* (Vol. 6, pp. 191–233). Greenwich, CT: JAI Press.

Feng, J., Spence, I., & Pratt, J. (2007). Playing an action video game reduces gender differences in spatial cognition. *Psychological Science, 18*(10), 850–855. doi:10.1111/j.1467-9280.2007.01990.x

Fiore, A. M., Kim, J., & Lee, -H.-H. (2005). Effect of image interactivity technology on consumer responses toward the online retailer. *Journal of Interactive Marketing, 19*(3), 38–53. doi:10.1002/dir.20042

Flowerday, T., & Schraw, G. (2003). Effect of choice on cognitive and affective engagement. *Journal of Educational Psychology, 96*(4), 207–215.

Fornell, C., & Larcker, D. (1981). Evaluating structural equation models with unobservable variables and measurement error. *Journal of Marketing Research, 18*, 39–50. doi:10.2307/3151312

Freeman, W. (2016, November). Does PlayStation 4 Pro really improve virtual reality performance? The Guardian. Retrieved from: https://www.theguardian.com/technology/2016/nov/14/does-playstation-4-pro-really-improve-virtual-reality-performance

Hayes, A. F. (2013). *Introduction to mediation, moderation, and conditional process analysis: A regression-based approach*. New York, NY: The Guilford Press.

Homer, P. M. (2006). Relationships among and ad-induced affect, beliefs, and attitudes: Another look. *Journal of Advertising, 35*(1), 35–51. doi:10.2753/JOA0091-3367350103

Horng, J.-S., & Tsai, C.-T. (2010). Government websites for promoting East Asian culinary tourism: A cross-national analysis. *Tourism Management, 31*, 74–85. doi:10.1016/j.tourman.2009.01.009

Hu, L., & Bentler, P. M. (1999). Cutoff criteria for fit indexes in covariance structure analysis: Conventional criteria versus new alternatives. *Structural Equation Modeling: A Multidisciplinary Journal, 6*(1), 1–55. doi:10.1080/10705519909540118

Huang, Y. C., Backman, K. F., Backman, S. J., & Chang, L. L. (2016). Exploring the implications of virtual reality technology in tourism marketing: An integrated research framework. *International Journal of Tourism Research, 18*(2), 116–128. doi:10.1002/jtr.v18.2

Hyun, M. Y., & O'Keefe, R. M. (2012). Virtual destination image: Testing a telepresence model. *Journal of Business Research, 65*, 29–35. doi:10.1016/j.jbusres.2011.07.011

Jacobs, C. (2004). *Interactive panoramas: Techniques for digital panoramic photography*. Berlin, NY: Springer.

Jennings, G., & Weiler, B. (2006). Mediating meaning: Perspectives on brokering quality tourism experiences. In G. Jennings & N. Nickerson (Eds), *Quality tourism experiences* (pp. 47–78). Burlington, MA: Elsevier Butterworth-Heinemann.

Jeong, C., Holland, S., Jun, S. H., & Gibson, H. (2012). Enhancing destination image through travel website information. *International Journal of Tourism Research, 14*, 16–27. doi:10.1002/jtr.v14.1

Jun, S. H., & Vogt, C. (2013). Travel information processing applying a dual-process model. *Annals of Tourism Research, 40*, 191–212. doi:10.1016/j.annals.2012.09.001

Jung, T., Chung, N., & Leue, M. C. (2015). The determinants of recommendations to use augmented reality technologies: The case of a Korean theme park. *Tourism Management, 49*, 75–86. doi:10.1016/j.tourman.2015.02.013

Kah, J. A., & Lee, S.-H. (2014). Beyond adoption of travel technology: Its application to unplanned travel behaviors. *Journal of Travel & Tourism Marketing, 31*(6), 667–680. doi:10.1080/10548408.2014.888968

Kang, M., & Gretzel, U. (2012). Effects of podcast tours on tourist experiences in a national park. *Tourism Management, 33*, 440–455. doi:10.1016/j.tourman.2011.05.005

Kerstetter, D., & Cho, M.-H. (2004). Prior knowledge, credibility and information search. *Annals of Tourism Research, 31*(4), 961–985. doi:10.1016/j.annals.2004.04.002

Kim, H., & Richardson, S. L. (2003). Motion picture impacts on destination images. *Annals of Tourism Research, 30*(1), 216–237. doi:10.1016/S0160-7383(02)00062-2

Klein, L. R. (2003). Creating virtual product experiences: The role of telepresence. *Journal of Interactive Marketing, 17*(1), 41–55. doi:10.1002/dir.10046

Kline, R. B. (2011). *Principles and practice of structural equation modeling*. New York: The Guilford Press.

Kuze, J., & Ukai, K. (2008). Subjective evaluation of visual fatigue caused by motion images. *Displays, 29*(2), 159–166. doi:10.1016/j.displa.2007.09.007

Lee, C.-K., Yoon, Y., & Lee, S. (2007). Investigating the relationships among perceived value, satisfaction, and recommendations: The case of the Korean DMZ. *Tourism Management, 28*(1), 204–214. doi:10.1016/j.tourman.2005.12.017

Lee, J.-S., Lee, C.-K., & Choi, Y. (2011). Examining the role of emotional and functional values in festival evaluation. *Journal of Travel Research, 50*(6), 685–696. doi:10.1177/0047287510385465

Lepp, A., Gibson, H., & Lane, C. (2011). Image and perceived risk: A study of Uganda and its official tourism website. *Tourism Management, 32*, 675–684. doi:10.1016/j.tourman.2010.05.024

López-Bonilla, J. M., & López-Bonilla, L. M. (2013). Self-service technology versus traditional service: Examining cognitive factors in the purchase of the airline ticket. *Journal of Travel & Tourism Management, 30*(5), 497–508. doi:10.1080/10548408.2013.803396

MacCannell, D. (1989). *The tourist: A new theory of the leisure class*. New York: Schochen.

Machleit, K. A., Allen, C. T., & Madden, T. J. (1993). The mature brand and brand interest: An alternative consequence of ad-evoked affect. *Journal of Marketing, 57*(4), 72–82. doi:10.2307/1252220

MacKay, K. J., & Fesenmaier, D. R. (1997). Pictorial element of destination in image formation. *Annals of Tourism Research, 24*(3), 537–565. doi:10.1016/S0160-7383(97)00011-X

McLuhan, M. (1964). *Understanding media: The extensions of man*. New York: McGraw Hill.

McMillan, S. J., & Hwang, J.-S. (2002). Measures of perceived interactivity: An exploration of the role of direction of communication, user control, and time in shaping perceptions of interactivity. *Journal of Advertising, 31*(3), 29–42. doi:10.1080/00913367.2002.10673674

Michaelidou, N., Siamagka, N. T., Moraes, C., & Micevski, M. (2013). Do marketers use visual representations of destinations that tourists value? Comparing visitors' image of a destination with marketer-controlled images online. *Journal of Travel Research, 52*(6), 1–16.

Okagaki, L., & Frensch, P. A. (1994). Effects of video game playing on measures of spatial performance: Gender effects in late adolescence. *Journal of Applied Developmental Psychology*, *15*(1), 33–58. doi:10.1016/0193-3973(94)90005-1

Reeves, B., & Nass, C. (1996). *The media equation: How people treat computers, television, and new media like real people and places*. New York: Cambridge University Press.

Reijnders, S. (2011). *Places of the imagination: Media, tourism, culture*. Burlington, VT: Ashgate Publishing Company.

Schraw, G., Flowerday, T., & Reisetter, M. (1998). The role of choice in reader engagement. *Journal of Educational Psychology*, *90*, 705–714. doi:10.1037/0022-0663.90.4.705

Sherry, J. L. (2004). Flow and media enjoyment. *Communication Theory*, *14*(4), 328–347. doi:10.1111/comt.2004.14.issue-4

Shiv, B., & Fedorikhin, A. (1999). Heart and mind in conflict: The interplay of affect and cognition in consumer decision making. *Journal of Consumer Research*, *26*(3), 278–292. doi:10.1086/jcr.1999.26.issue-3

Suh, K.-S., & Chang, S. (2006). User interfaces and consumer perceptions of online stores: The role of telepresence. *Behaviour & Information Technology*, *25*(2), 99–113. doi:10.1080/01449290500330398

Sundar, S. S. (2000). Multimedia effects on processing and perception of online news: A study of picture, audio, and video downloads. *Journalism and Mass Communication Quarterly*, *77*(3), 480–499. doi:10.1177/107769900007700302

Sundar, S. S. (2008). The MAIN model: A heuristic approach to understanding technology effects on credibility. In M. J. Metzger & A. J. Flanagin (Eds.), *Digital media, youth, and credibility* (pp. 72–100). Cambridge, MA: The MIT Press.

Sundar, S. S. (2009). Media effects 2.0: Social and psychological effects of communication technologies. In R. L. Nabi & M. B. Oliver (Eds.), *The Sage handbook of media processes and effects* (pp. 545–560). Thousand Oaks, CA: Sage Publications.

Sundar, S. S., Narayan, S., Obregon, R., & Uppal, C. (1998). Does web advertising work? Memory for print vs. online media. *Journalism & Mass Communication Quarterly*, *75*(4), 822–835. doi:10.1177/107769909807500414

Sundar, S. S., Xu, Q., & Dou, X. (2012). Role of technology in online persuasion: A MAIN model perspective. In S. Rodgers & E. Thorson (Eds.), *Advertising theory* (pp. 355–372). New York: Routledge.

Tabachnick, B. G., & Fidell, L. S. (2000). *Computer-assisted research design and analysis*. Needham Heights, MA: Allyn & Bacon.

Tang, L., & Jang, S. (2012). Investigating the routes of communication on destination websites. *Journal of Travel Research*, *51*(1), 94–108. doi:10.1177/0047287510394196

Tussyadiah, I. P., & Fesenmaier, D. R. (2009). Mediating tourist experiences: Access to places via shared videos. *Annals of Tourism Research*, *36*(1), 24–40. doi:10.1016/j.annals.2008.10.001

Urry, J. (2000). *Sociology beyond societies*. London: Routledge.

Urry, J. (2002). *The tourist gaze*. London, UK: Sage Publications Ltd.

Vorderer, P., Wirth, W., Saari, T., Gouveia, F. R., Biocca, F., JäNcke, F., ... JäNcke, P. (2003). *Constructing presence: Towards a two-level model of the formation of spatial presence* (No. IST-2001- 37661). Unpublished report to the European Community, Society & Technology (IST) Program, Project Presence: MEC. Hannover, München, Helsinki, Porto, Zurich.

Wearing, S., Stevenson, D., & Young, T. (2010). *Tourist cultures: Identity, place, and traveller*. Thousand Oaks, CA: Sage Publications.

Wirth, W., Hartmann, T., Böcking, S., Vorderer, P., Klimmt, C., Schramm, H., ... Jäncke, P. (2007). A process model of the formation of spatial presence experiences. *Media Psychology*, *9*(3), 493–525. doi:10.1080/15213260701283079

Wu, G. (2010). The mediating role of perceived interactivity in the effect of actual interactivity on attitude toward the website. *Journal of Interactive Advertising*, *5*(2), 29–39. doi:10.1080/15252019.2005.10722099

Wu, J. J., & Chang, Y. S. (2005). Effects of transaction trust on e-commerce relationships between travel agencies. *Tourism Management*, *27*(6), 1253–1261. doi:10.1016/j.tourman.2005.06.009

Zeithaml, V., Berry, L., & Parasuraman, A. (1996). The behavioral consequences of service quality. *The Journal of Marketing*, *60*(2), 31–46. doi:10.2307/1251929

Zhang, B., Appelman, A., Choi, Y., & Han, A. (2013). The effect of interactivity on user experience: Comparing visual cues and actual use. *Paper presented at the 2013 Conference of the International Communication Association, London, U.K.*

PART 2
Visuals and film tourism

Perceived values of TV drama, audience involvement, and behavioral intention in film tourism

Seongseop (Sam) Kim and Sangkyun Kim

ABSTRACT

This study attempts to develop a drama consumption model in the context of film tourism and empirically test it using the mainland Chinese audience of Korean television dramas (K-dramas). Celebrity and dramatized characters strongly influenced the emotional involvement and referential reflection of this audience, whereas the effects of filming location and backdrops were relatively weak in this respect. It is worth noting that the perceived values of K-dramas did not have a direct effect on behavioral intention to visit film tourism locations. Rather, they are expected to lead to film tourism through a process of psychological and emotional involvement with the dramas. Further studies can beneficially apply this model to measure media value, audience involvement, and intention in different social and geographical settings.

Introduction

Film tourism is one of the fastest growing niche tourism areas, and has received an increasing amount of attention from scholars in various academic disciplines over the last decade or so (Connell, 2012; Kim & Long, 2012). Existing academic knowledge about film tourism is, however, still highly fragmented and mostly focuses on destination marketing and a management perspective, often leading to narrowly focused findings that fail to provide a holistic understanding of film tourism impacts and their broader implications. While this narrow approach has enhanced our understanding of destination image enhancement, destination choice, and film tourist motivations, more recent film tourism studies have been keenly interested in assessing the underlying mechanisms of the film tourism phenomenon by adopting various theoretical concepts such as cultural proximity (see for example Su, Huang, Brodowsky, & Kim, 2011), celebrity worship (Yen & Croy, in press), destination image (see for example Yen & Croy, in press), perceived values of media attributes (see for example Kim, 2012a, 2012b), motivation (see for example Rittichainuwat & Rattanaphinanchai, 2015), authenticity (see for example Buchmann, Moore, & Fisher, 2010), nostalgia (see for example Reijnders, in press), and place attachment (see for example Wong & Lai, 2015). Although each study has proposed a different set of research objectives in its own research context, the common denominator, to a greater or lesser extent, is the nature of film tourism, which has "a close relationship with visual, aural, and textual media representations, their aesthetic values and decoded symbolic meanings behind the screen through media production and consumption" (Kim & Assaker, 2014, p. 252). Without understanding this connection in the film tourism context, it is difficult to understand the antecedents and consequences of film tourists' motivations, behavioral intentions, and their actual on-site experiences.

Thus, media production values and audience involvement have been considered to be two of the major concepts to enrich our understanding of film tourism from the experiential perspective. However, only a few empirical studies using quantitative methodologies have examined the close relationships between celebrity involvement and behavioral intention, as well as their moderating or mediating role in predicting those relationships in the context of film tourism (see for example Wong & Lai, 2015; Yen & Croy, in press; Yen & Teng, 2015).

Nonetheless, no previous film tourism studies have empirically examined the structural relationships of the multidimensional audience involvement concept, which plays a crucial role in understanding audiences' actual viewing experiences of media programs and influences an audience's behavioral intention to visit film tourism locations. This study examines the relationships between the values of TV dramas, audience involvement, and behavioral intention by adopting structural equation modeling (SEM) methodologies with a data sample of mainland Chinese audiences of K-dramas. Through these efforts, we expect to improve our current understanding of the close relationships between visual, aural, and textual media representations, their values, and the decoded symbolic meanings behind the screen in the process of media production and consumption, in addition to their association with actual film tourism.

Literature review and hypotheses

Values of TV drama, audience involvement, and behavioral intention

Production values have been defined as "the professional appearance, or polish, of a production, including video and audio quality (e.g. visual technique, televisualized esthetics, music, and sound effects), lighting, the number of errors, and the amount and quality of special effects (spectacle as visual attractions) in TV and film production" (Kim & O'Connor, 2011, p. 245). Similarly, Bae and Lee (2004) conceptually and empirically suggested that media factors equivalent to the concept of production values – including media type, genre, topic, characters and stars, depiction of interpersonal relationships, and visual techniques (e.g. close-up shots) – are direct antecedents of audience involvement. Sood and Rogers (2000) specifically highlighted the ways in which different media technologies evoke different levels of audience involvement.

The concept of audience involvement is rooted in "parasocial interaction" (Giles, 2002; Horton & Wohl, 1956) and refers to an imagined connectedness between an individual audience member and a media character (and often a celebrity). It has been frequently documented in previous film tourism studies under different terms, including vicarious or empathic response (Beeton, 2005; Frost, 2010; Reijnders, 2010; Riley, Baker, & Van Doren, 1998; Yen & Croy, in press). These film tourism studies have highlighted the theoretical significance of vicarious, emotional, and empathic audience involvement to enable the researchers to better understand the film tourism phenomenon without empirical tests. Some recent studies have attempted to empirically examine the role of audience involvement concept in constructing on-site film tourist experiences (Kim, 2012a; Kim & Assaker, 2014). Other studies have specifically tested celebrity-focused involvement or engagement in relation to potential tourists' destination image perceptions of locations depicted in films or TV dramas, and their subsequent behavioral intention (Lee, Scott, & Kim, 2008; Wong & Lai, 2015; Yen & Croy, in press; Yen & Teng, 2015).

Building on the aforementioned studies, and as the first attempt of its kind, Kim and Assaker (2014) used an SEM model to empirically examine the structural relationships between production values and audience involvement in Korea in the context of film tourism associated with the Korean historical TV drama, *Daejanggeum*. The direct, positive effect of production values on audience involvement was confirmed, suggesting that the production values had two dimensions (story and character, and location and visual effects), whereas audience involvement constituted three dimensions (behavioral involvement, emotional involvement, and referential reflection). Hence, the following hypotheses (Hs) are proposed:

H1: The values of TV dramas have a direct positive effect on the audience's level of emotional involvement.

H2: The values of TV dramas have a direct positive effect on the audience's level of referential reflection.

H3: The values of TV dramas have a direct positive effect on the level of the audience's behavioral involvement.

Furthermore, it is theoretically and conceptually suggested that the perceived production values of a TV drama or film become a main motivational driver that results in an audience member's behavioral intention to be a film tourist (Frost, 2010; Macionis & Sparks, 2009; Rittichainuwat & Rattanaphinanchai, 2015; Yen & Croy, in press). As proposed by parasocial theory (Giles, 2002), characters or important representations in TV dramas cause the audience to feel closer to the film locations, and repeated reciprocal interactions with them will contribute to the increasingly higher likelihood of intention to participate in film tourism. Thus, the following hypothesis is proposed:

H4: The values of TV dramas have a direct positive effect on the audience's behavioral intention to participate in film tourism.

Structural relationships between emotional involvement, behavioral involvement, and referential reflection

No previous film tourism studies have empirically examined the structural relationships to identify the multidimensional value of audience involvement. However, in the field of media and communication studies, Sood (2002) not only theoretically confirmed its five dimensions, but also statistically proved their structural relationships, conceptualizing audience involvement as the degree to which audience members engage in reflection upon, and parasocial interaction with, certain media programs, thus resulting in overt behavior change. Audience involvement can be seen as being composed of two main elements: (1) reflection (critical and/or referential), and (2) parasocial interaction (cognitive, affective, behavioral participation or any combination of these) with the media (Sood, 2002, p. 156).

Some studies have also affirmed the importance of the affective and emotional dimensions of audience involvement (i.e. referential reflection and emotional interaction) in explaining the cognitive and behavioral dimensions of audience involvement (Kim, 2012a; Sood, 2002). Both referential reflection and emotional involvement are grounded in the affective and emotional psychology of an audience's engagement with a media program. However, the cognitive dimension of audience involvement did not seem to be significant in the context of film tourism in earlier exploratory studies (Kim, 2012a; Kim & Assaker, 2014). This is understandable, given the fact that other previous film tourism studies collectively adopted and tested the affective and emotional aspect of an audience's involvement using vicarious involvement, empathic involvement, and celebrity involvement or worship (Lee et al., 2008; Wong & Lai, 2015; Yen & Croy, in press; Yen & Teng, 2015).

Thus, individuals who engage in high levels of referential and affective involvement with a media program are more likely to talk to, or about, media characters and celebrities during and after exposure, and further communicate interpersonally with their friends, colleagues, and family members as well as anonymous virtual community members (Lee et al., 2008; Papa et al., 2000; Sood, 2002; Sood & Rogers, 2000), which are the characteristics of behavioral involvement. The following hypotheses are therefore proposed:

H5: The level of the audience's emotional involvement has a direct positive effect on its behavioral involvement.

H6: The level of the audience's referential reflection has a positive effect on its behavioral involvement.

When an audience member develops a deep emotional connection with a media program, he or she also forms a referential reflection. Referential reflection refers to the degree to which an individual relates a media program to his or her personal experiences (Sood, 2002), and connects the program to the viewer's real life in terms of his or her own life and problems, as if he or she were relating to the media characters as real people and, in turn, relating these real people to his or her own real world (Roesch, 2009; Sood & Rogers, 2000). In this respect, Sood (2002) found that when audiences exhibited a strong identification with the characters and celebrities (i.e. emotional involvement), they also related the media program to their personal lives and stories more actively (i.e. referential reflection). This is because emotional involvement, through the development of parasocial interaction, identification, empathy, and/or reflection, is at the center of the audience's viewing experience, in particular in the genre of serialized TV dramas (Kim & Long, 2012). Based on the above discussion, the following hypothesis is proposed:

H7: The level of the audience's emotional involvement has a direct positive association with its referential reflection.

Audience involvement and behavioral intention

The degree of audience involvement influences the degree of pleasure in viewing experiences, subsequent planned media exposure, levels of viewing experience satisfaction, and post-viewing behavior intention, including purchasing behavior (Bae & Lee, 2004; Sood, 2002). It extends to the context of film tourism, in which the more emotional and behavioral involvement. The more referential reflection audiences develop through viewing a TV drama, the greater the likelihood of the audience visiting film tourism locations and the greater the satisfaction they feel when doing so (Carl, Kindon, & Smith, 2007; Kim, 2012a; Kim & Assaker, 2014; Yen & Teng, 2015).

From the film tourism perspective, the affective element of audience involvement, along with behavioral involvement, plays a critical role in forming personalized memories and symbolic meanings, creating new tourism spaces, and contextualizing the anticipation of tourist experiences at film locations, of which the latter is the consequence of behavioral intentions to visit the film locations (Kim & Assaker, 2014; Lee et al., 2008; Reijnders, 2010). In this respect, some suggest that a positively perceived destination image of filmed locations arising from the development of a bond between a viewer and media character(s) and content(s) while watching a media program is an immediate antecedent of behavioral intention (Wong & Lai, 2015; Yen & Croy, in press; Yen & Teng, 2015). Hence, the following hypotheses are proposed:

H8: The level of the audience's emotional involvement has a positive effect on the level of behavioral intention to participate in film tourism.

H9: The level of the audience's referential reflection has a positive effect on the level of behavioral intention to participate in film tourism.

H10: The level of the audience's behavioral involvement has a positive effect on the level of behavioral intention to participate in film tourism.

Figure 1 illustrates which of the paths between concepts are supported and illustrates which of the paths between concepts are supported and validate the TV drama consumption model in the context of film tourism. This proposed TV drama consumption model is differentiated from the aforementioned previous studies in that there was no empirical examination to support the causal relationships between the perceived values of TV drama, audience involvement, and behavioral intention. As a result of this, the complex decision-making process of potential film tourists (i.e. behavioral intention), intertwined with personal assessments of TV drama production values and levels of audience involvement developed prior to the actual visit, were not satisfactorily tested and confirmed. The work of Kim and Assaker (2014) mainly focused on the antecedents of on-site film tourism experiences, which take place in the during-visiting stage, rather than in the pre-visit stage, dealing with intention to actual visit. Thus, this conceptual model and its subsequent empirical confirmation in this study will fill the currently existing gaps mentioned above.

Research method

Study setting

The popularity of K-dramas has been noted since the 1990s, but recently *My Love From the Star*, a 20-episode K-drama that was released in China in the spring of 2014, after its initial airing in Korea in December 2013, reignited China's craze for them. The online viewership of this K-drama reached over 2.5 billion in mainland China, making it the most watched of all TV programs during the nine weeks of its airing in 2014 (Zhang & Choi, 2015). This had a spillover effect on retail industries such as beauty and food. For example, in one scene the main actress was shown eating Korean fried chicken and drinking beer, which is one of the most common snacks in Korea (the so-called "chimaek", a

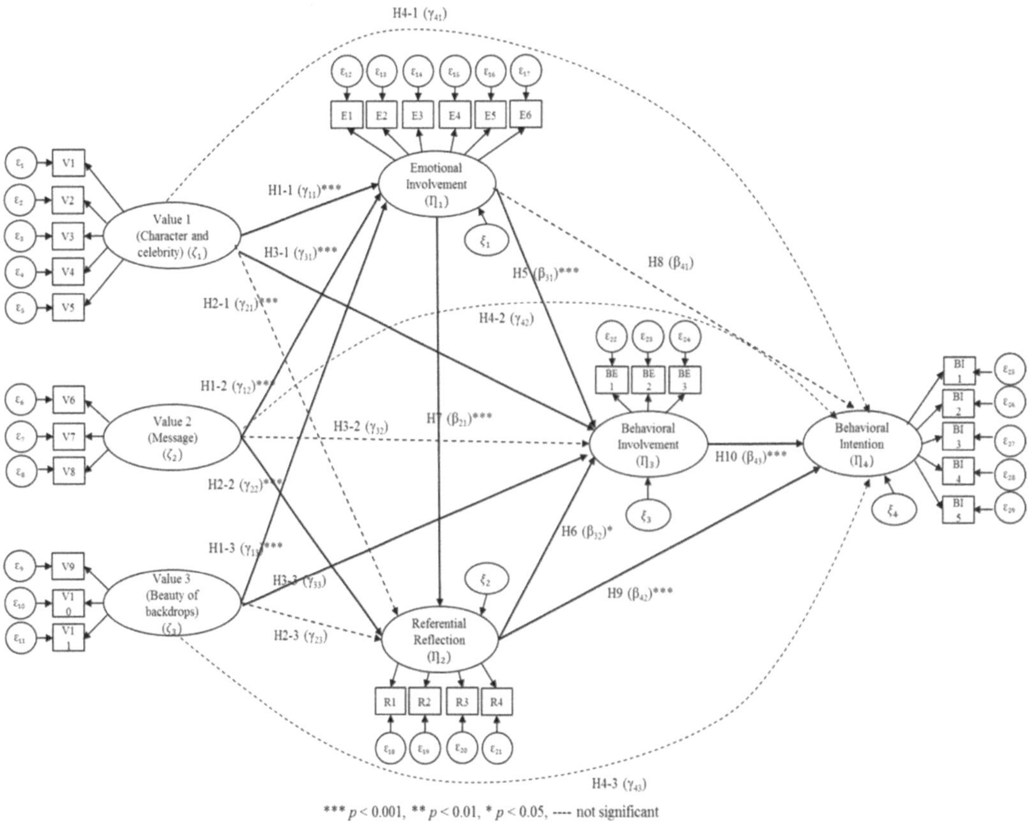

Figure 1. Relationship paths.

portmanteau for chicken and beer in the Korean language). With its appearance, in-bound tourists, as a result of viewing *My Love From the Star,* have made "pilgrimages" to Korean fried chicken franchise restaurants. "Chimaek" has become a must-have item in Korea, and one Korean fried chicken franchise is planning to increase the number of franchisees to 10,000 by 2020 (Lee, 2015). In March 2016, about 4500 company workers visited filming locations of the drama and ate food which was depicted in it (Chosun Ilbo, 2016).

Measurement

The development of items for the perceived values of K-dramas was conducted by reviewing previous studies (see for example Frost, 2010; Kim, Agrusa, Lee, & Chon, 2007; Kim & Assaker, 2014; Kim & Long, 2012; Kim & O'Connor, 2011; Macionis & Sparks, 2009; Riley et al., 1998; Roesch, 2009; Yen & Teng, 2015). The items include the beauty of backdrops, performance of star actors and their influence, storyline, props, embedded meaning or educational message of the media content, and psychological attachment to the shooting locations. The measurement items for audience involvement were designed to measure emotional involvement, referential reflection, and behavioral involvement, as in previous studies (Kim, 2012a; Kim & Assaker, 2014; Lee et al., 2008; Sood, 2002; Yen & Croy, in press). The items used to measure behavioral involvement were designed to indicate behavioral loyalty to K-dramas.

Along with the development of items based on the literature, in-depth interviews were conducted with two groups of five mainland Chinese graduate students who like K-dramas. Over the course of two meetings, they revealed features of K-dramas, differences between them and those of mainland Chinese TV dramas, their responses to and actions regarding the K-dramas after watching them, and their behavioral intention to seek further experiences. Examples of items added through the qualitative research were pertinent to behavioral intention, such as "I'd like to buy a Korean consumer product (e.g. an electronic device or cosmetics) on a tour in Korea" and "I'd like to buy products displayed on screen or accessories put on by star actors/actresses during dramas". All of the aforementioned items were measured on a 5-point Likert scale in which "1" equals "strongly disagree" and "5" equals "strongly agree".

In developing or extracting items, it is important to alleviate the concern of common method biases, which are attributed to "measurement methods rather than the constructs the measures represent" (Bagozzi & Yi, 1991). Here, method indicates the form of measurement at different levels of abstraction, such as the specific items, scale type, response format, and the general context (Bagozzi & Yi, 1991). Podsakoff and MacKenzie (2003) also summarized a list of potential sources of common method biases, including item characteristic effects (e.g. social desirability, item ambiguity, item demand characteristics), item context effects (e.g. item priming effects, item embeddedness, context-induced mood, scale length, intermixing of items), common rater effects (e.g. consistency motif, transient mood state), and measurement context effects (e.g. independent variables and dependent variables measured at the same point in time, in the same location).

In order to minimize concerns regarding these common method biases, the alignment or wording of the items was meticulously reviewed. For example, some questions were reversely worded to avoid consistency motif bias, whereas all questions were designed to ask about actual perceptions rather than social trends. To alleviate transient mood state bias, a main survey was administered two or three months after the end of the TV drama. The questions did not hide any cues designed to induce intentional answers, in order to avoid item demand characteristics bias. Where possible, positive and negative items were aligned with mixture, due to the concern regarding positive and negative item bias. To reduce item context effects bias, the first questions were designed to ask about the participants' interest in K-drama or demographic variables, whereas items to test causal relationships were separately arranged. In addition to this, this study mitigated measurement context effect bias, which can occur in face-to-face interview surveys, interview bias, and regional sampling (Podsakoff & MacKenzie, 2003), by employing online panelist data two and three months respectively after the TV drama had ended; the panelists were also selected from all over China.

In addition to this, the testing of reliability and validity in the process of item development and before main data analyses was performed and is related to concerns regarding common method biases because reliability and reliability tests are designed to identify both if the measurement is of quality in repeated observations of the same phenomenon and if the measurement is accurate (Babbie, 1995). This study introduced reliability and a variety of validity of constructs through scale development and during data analyses. The SEM model, which consists of seven constructs or dimensions related to the TV drama consumption model, needs to check validity for the scale, which indicates the extent to which the concept was precisely measured. Three major types of validity include content validity, criterion-related validity, and construct validity (Babbie, 1995).

First, content validity can be achieved through a thorough literature review, pre-test, and pilot test. This study involved a rigorous literature review, pre-test, and pilot-test. In particular, this study checked face validity using a pool of judges (professors and graduate students) during the pre-test, to guarantee a comprehensive understanding of the model comprising several constructs.

In order to confirm criterion-related validity, this study measured predictive validity, which is the extent to which a score on a scale or test predicts scores in some criterion (Babbie, 1995). This study also used correlation analysis to ascertain concurrent validity among the constructs. That is, a high level of correlation between similar constructs ensures satisfactory concurrent validity (Hair, Anderson, Tatham, & Black, 2009). In addition to this, confirmatory factor analysis (CFA) using the main survey data set will be assessed, to check construct validity and discriminant validity.

A questionnaire was developed based on the literature review and in-depth interviews. The questionnaire in English was then translated into Chinese by three PhD students who were majoring in tourism and hospitality. The translators met twice to exchange their comments on the translated content and then a final version was decided upon. A pilot test was subsequently conducted in the form of an online panel survey, using a sample of 50 mainland Chinese people. The pilot test respondents completed the questionnaire without substantial concerns.

The main survey was thus conducted using an online panel survey method, which has certain advantages such as a rapid response rate through the Internet, the possibility of selecting exact target samples, and timely answers from respondents (Grönlund & Strandberg, 2014). One restriction placed on the selection of the target sample was that respondents must have watched at least two K-dramas since 2013. The data were collected from June 5 to June 25, 2014, using an online panel company with approximately 10 million members in China. Of 450 questionnaires completed, 20 including doubtful or incomplete answers were excluded, and 430 completed questionnaires were used for further data analysis.

Table 1. Demographic profiles (N = 430).

Variable	Category	Frequency	Percentage (%)
Gender	Male	135	31.4
	Female	295	68.6
Age (years)	20–25	107	24.9
	26–30	159	37.0
	31–35	99	23.0
	36–40	37	8.6
	41 or above	27	6.3
Marital status	Single	175	40.7
	Married	253	58.8
	Other	2	0.5
Education	High school or below	12	2.8
	College student	7	1.6
	College	411	95.6
Occupation	Company employee	243	56.5
	Own business	12	2.8
	Civil servant	16	3.7
	Agricultural/Fishery	1	0.2
	Professional	30	7.0
	Housewife	4	0.9
	Technician	41	9.5
	Student	42	9.8
	Sales/Service employee	14	3.3
	Teacher	20	4.7
	Driver/Transportation	3	0.7
	Other	4	0.9

Results

Demographic profile

As Table 1 shows, nearly 69% of the respondents were female and almost 59% were married. The great majority of the respondents (almost 96%) were college graduates or above and nearly 57% were company employees. Here, higher percentages of females, young people, and company workers may lead to concerns regarding sampling bias. However, according to a report on mainland Chinese fandom, the main audience of K-dramas is highly educated, young, and female, and consists of Internet users (Zhang & Choi, 2015), which helps to mitigate the worry regarding representativeness of the sample. Table 2 shows that more than 78% of the respondents indicated that they were somewhat or strongly interested in K-dramas. Watching K-dramas via the Internet was the most popular method of viewing (75.3%), followed by watching on television (20.2%). More than 88% of respondents indicated that they watched K-dramas for an average of over one hour per week. Of these, nearly 31% indicated that the average amount of time spent watching K-dramas was more than five hours per week. About 90% of the respondents reported satisfaction with K-dramas. Considering the image change regarding Korea after watching K-dramas, 79% of respondents reported that their perception of Korea had been positively changed.

Exploratory factor analyses and reliability tests

In order to extract the underlying domains of perceived values of K-dramas, principal components and varimax rotation methods were used. As shown in Table 3, the factor analysis, using 11 items, generated a three-factor model in which the items have an eigenvalue greater than 1.0. The factor structure explained 60.91% of the variance. Factor loading the measure correlation between the observed measurements and the factors also exceeded Comrey and Lee's (1992) criterion (0.63), indicating a "very good" factor model. All reliability

Table 2. Characteristics of respondents (N = 430).

Variable category		Frequency	Percentage (%)
Interest in K-dramas	Somewhat disagree	3	0.7
	Neutral	89	20.7
	Somewhat agree	244	56.7
	Strongly agree	94	21.9
Method of watching K-dramas	TV	87	20.2
	Satellite TV channel	11	2.6
	Smartphone	8	1.9
	Internet	324	75.3
Frequency of watching K-dramas (per week)	None	1	0.2
	Less than 30 minutes	19	4.4
	30 minutes–1 hour	30	7.0
	1.1–3 hours	106	24.7
	3.1–5 hours	141	32.8
	5.1 hours or more	133	30.9
Level of satisfaction	Strongly dissatisfied	0	0
	Somewhat dissatisfied	4	0.9
	Neutral	40	9.3
	Somewhat satisfied	305	70.9
	Strongly satisfied	81	18.8
Change of perceptions of Korea since watching K-dramas	Very unfavorable	0	0
	Somewhat unfavorable	3	0.7
	Same	88	20.5
	Somewhat favorable	270	62.8
	Very favorable	69	16.0

Table 3. Exploratory factor analysis and reliability tests for values of K-dramas.

Domain	Item	Factor loading	Mean	S.D.
Value 1 (Character and celebrity. 4.41 [a]; 40.09 [b]; .78 [c])	Enhanced emotional involvement and identification (e.g. close-up scenes capturing characters and icons)	.71	4.20	.65
	Attractiveness of the main characters (e.g. ways of speaking, behavior, personality)	.75	4.22	.67
	Casting overall (e.g. casting and imaginative content)	.60	4.17	.63
	Interesting content (e.g. creative and imaginative content)	.65	4.17	.65
	Celebrity/star power (e.g. reputation, fame, popularity)	.69	4.29	.66
Value 2 (Message. 1.28; 11.64; .75)	Educational and moral messages (e.g. humanism, friendship, true love)	.79	3.71	.82
	Embedded meanings in certain parts of the set	.88	3.63	.80
	Topicality (topics and themes that do not cause offence to my country's culture, religion, or relations with my country)	.64	3.81	.72
Value 3 (Beauty of backdrops. 1.01; 9.18; .73)	Spectacle (e.g. props from drama, costumes, furniture, accessories, special effects)	.68	4.17	.71
	Beautiful scenic background (e.g. natural scenery)	.82	4.27	.71
	Personal attachment to locations in the drama (depiction of places that appeal to me)	.75	4.04	.71

[a]eigenvalue; [b]variance; [c]reliability alpha.

alpha scores within the three domains were greater than 0.7, indicating internal consistency of the items within each factor (Nunnally, 1978). The extracted domains were labeled "character and celebrity", "message", and "beauty of backdrops". The mean values for the 11 items ranged from 3.63 to 4.29, and the grand means for the three domains were 4.21, 3.71, and 4.16 respectively.

A series of exploratory factor analyses for emotional involvement (six items), referential reflection (four items), behavioral involvement (three items), and behavioral intention (five items) using principal components and varimax rotation methods generated single-factor models. Each construct explained 64.41, 74.16, 71.58, and 68.71% of the variance respectively. Factor loadings on all four constructs were greater than .74, which exceeds Comrey and Lee's (1992) criterion (0.71) for an "excellent" factor model. The reliability alphas were .89, 88, 80, and 91 respectively, indicating internal consistency among the items of each construct. The results are reported in Table 4. Table 5 shows the results of the correlation analysis. Significances at the .001 level were observed between all pairs of constructs. Thus, concurrent validity was confirmed.

In order to test the predictive validity of the underlying domains of the perceived value of K-dramas, multiple regression analyses were conducted to predict dependent variables, such as the levels of satisfaction and image changes after watching. If a high explanatory power (R^2) is found, predictive validity is ascertained (Babbie, 1995). Since the three domains of TV

Table 4. Exploratory factor analyses and reliability tests for emotional involvement, referential reflection, behavioral involvement, and behavioral intention.

Construct	Item	Factor loading	Mean	S.D.
Emotional involvement (3.86[a]; 64.41[b]; .89[c])	I feel that the K-dramas and my favorite character(s)/actor(s) keep me company.	.84	3.37	.97
	I like hearing the voice(s) of my favorite character(s)/actor(s) in my home.	.84	3.70	.84
	I feel comfortable when watching K-dramas, as if the character(s)/actor(s) were my friends.	.79	3.79	.80
	I really miss K-dramas and my favorite character(s)/actor(s) whenever I am unable to watch them.	.79	3.77	.96
	My favorite characters/actors in K-dramas are like old friends.	78	3.83	.81
	When watching K-dramas, I feel that I am a part of the story.	.77	3.53	.91
Referential reflection (2.97; 74.16; .88)	K-dramas seem to understand and cover the kinds of issues that I want to know more about.	.89	3.54	.89
	I feel that K-dramas portray real life lessons and practices that I can personally relate to.	.86	3.62	.77
	When character(s)/actor(s) in K-dramas express an opinion, it helps me make up my own mind about the issue.	.86	3.68	.78
	I feel that K-dramas present things as they really are in life.	.84	3.55	.98
Behavioral involvement (2.15; 71.58; .80)	If my favorite character(s)/actor(s) appear on another program, I watch that program.	.85	4.03	.79
	If there is a story about K-dramas or my favorite character(s)/actor(s) in a newspaper, magazine or on the Internet, I read it.	.85	4.03	.74
	I look forward to watching K-dramas and my favorite character(s)/actor(s) in each episode.	.84	4.17	.68
Behavioral intention (4.12; 68.71; .91)	I'd like to visit a Korean restaurant that has been shown in a TV drama.	.88	3.34	1.05
	I'd like to participate in a fan meeting in Korea if the opportunity is given.	.86	3.54	1.00
	I'd like to buy Korean foods that have been featured in a TV drama.	.86	3.43	.96
	I'd like to buy a Korean consumer product (e.g. electronic device, cosmetics) on a tour in Korea.	.76	3.81	.94
	I'd like to visit K-drama shooting locations someday.	.74	3.95	.85

[a]eigenvalue; [b]variance; [c]reliability alpha.

Table 5. Correlations among latent constructs (squared).

	Constructs						
	V1	V2	V3	EI	RR	BI	IT
V1	1.00						
V2	0.583 (0.340)	1.00					
V3	0.727 (0.529)	0.537 (0.288)	1.00				
EI	0.563 (0.317)	0.752 (0.566)	0.595 (0.354)	1.00			
RR	0.446 (0.199)	0.854 (0.729)	0.472 (0.223)	0.790 (0.624)	1.00		
BI	0.750 (0.563)	0.585 (0.342)	0.646 (0.417)	0.763 (0.582)	0.606 (0.367)	1.00	
IT	0.533 (0.284)	0.639 (0.408)	0.498 (0.248)	0.837 (0.701)	0.691 (0.477)	0.735 (0.540)	1.00

V1: character and celebrity; V2: message; V3: beauty of backdrops; EI: emotional involvement; RR: referential reflection; BI: behavioral involvement; IT: behavioral intention to be a film tourist.
All correlations ($p < 0.001$).

drama values explained 37% and 29% of the variance of the two dependent variables respectively, predictive validity was satisfactory.

Confirmatory factor analysis

Before conducting the SEM procedure, a confirmatory factor analysis (CFA) was performed to guarantee that the proposed measurement model specified the posited relations of the observed variables to the latent constructs. As the results of the CFA indicated construct reliability, convergent validity, and discriminant validity, the statistical approach was appropriate, compared with an exploratory factor analysis, for producing a factor structure. The maximum likelihood method of estimation was used to test the overall conceptual model. The results of the CFA revealed an acceptable model fit, except for the chi-square value ($\chi^2(384) = 803.74$, $p = .000$; TLI = 0.94; CFI = 0.94; RMSEA = 0.05; NFI = 0.90; GFI = 0.88; AGFI = 0.86). However, since the chi-square value is sensitive to sample size, the other model fit indices are considered to substantially evaluate the model.

A measurement model is considered reliable when the correlated component regression (CCR) values are greater than 0.7 (Hair et al., 2009). As all seven constructs had CCR values greater than the criterion (ranging from 0.84 to 0.92), the construct validity was considered to be satisfactory. Validity was tested with the scale of *t*-values for each construct and a latent variable, and the average variance extracted (AVE). All 29 items in the measurement model had a significant *t*-value for the latent construct. This indicates that the measurement scales for each construct support high levels of convergence and construct validity. In addition, the AVE values in each construct were greater than 0.5, indicating a high level of convergent validity.

Discriminant validity shows the extent to which a given construct is different from other constructs (Hair et al., 2009). To evaluate discriminant validity, the AVE values for each construct should be greater than the squared correlations between the construct and all other constructs (Fornell & Larcker, 1981). The AVE values ranged from 0.63 to 0.72 and the highest squared correlations were 0.73 and 0.70. As the squared correlations were close to or lower than the AVE values, the measurement model demonstrates discriminant validity.

SEM

In order to ascertain whether the hypothesized K-drama consumption model was consistent with the measured data, an empirical SEM with the maximum likelihood of estimation was assessed. The covariance matrix (shown in Table 5) was input, to further test the hypothesized conceptual model. Table 6 presents the results of the SEM analysis, including goodness-of-fit indices and all relation paths, based on the 18 hypotheses. If a path between two constructs is positively significant at the level of .05, this indicates that one exogenous variable has a positive effect on an endogenous variable.

In terms of model fit indices, the overall data set was satisfactory, except for the chi-square value, which was statistically significant ($\chi^2(386) = 1033.38$, $p = .000$; TLI = 0.90; CFI = 0.91; RMSEA = 0.06; NFI = 0.87; GFI = 0.86; AGFI = 0.83). Next, the path coefficients of the 18 relation paths were estimated. Table 6 and Figure 1 illustrate the overall results of the conceptual model.

The results of the SEM designed to test hypotheses 1–1, 1–2, and 1–3 showed that all three paths were significant at the .001 level. Paths were identified between "character and celebrity" and "emotional involvement" ($\gamma_{11} = .27$, $t = 3.70$, $p < 0.001$), between "message" and "emotional involvement" ($\gamma_{12} = .90$ $t = 8.55$, $p < 0.001$), and between "beauty of backdrops" and "emotional involvement" ($\gamma_{13} = .33$, $t = 5.42$, $p < 0.001$). Thus, these three hypotheses were supported. Tests of hypotheses 2–1, 2–2, and 2–3 showed that paths between "message" and "referential reflection" were significant at the .001 level ($\gamma_{22} = .84$, $t = 7.60$, $p < 0.001$). Hypothesis 2–2 was thus supported.

In tests of hypotheses 3–1, 3–2, and 3–3, only the path between "character and celebrity" and "behavioral involvement" ($\gamma_{31} = .48$, $t = 7.14$, $p < 0.001$), and "beauty of backdrops" and "behavioral involvement" ($\gamma_{33} = .16$, $t = 3.45$, $p < 0.001$) were found to be significant at the .001 level; hypotheses 3–1 and 3–3 were thus supported. The tests of hypotheses 4–1, 4–2, and 4–3 were not significant even at the .05 level. However, a path from "emotional involvement" to "behavioral involvement" was positively significant at the .001 level ($\beta_{31} = .29$, $t = 4.19$, $p < 0.001$). This confirmed that hypothesis 5 was supported.

"Referential reflection" was positively related to "behavioral involvement" at the .001 level, so hypothesis 6 was also supported ($\beta_{32} = .17$, $t = 1.960$, $p < 0.05$). Testing hypothesis 7 revealed that "emotional involvement" has a positive effect on "referential reflection" ($\beta_{21} = .39$, $t = 5.68$, $p < 0.001$). On testing hypothesis 8, "emotional involvement" was found to have no effect on "behavioral intention," but hypothesis 9 was supported as a positive relationship was found between "referential reflection" and "behavioral intention" ($\beta_{42} = .68$, $t = 6.24$, $p < 0.001$). Finally, "behavioral involvement" was found to have a positive effect on "behavioral intention" ($\beta_{43} = .60$, $t = 4.44$, $p < 0.001$); hypothesis 10 was thus supported. The path relations are reported in Figure 1.

Conclusion and discussion

Based on the above results of the SEM analysis, the key findings of this study are as follows. First, characters and celebrities in TV drama were influential in explaining all three underlying dimensions of audience involvement. This is consistent with previous studies highlighting the fact that favorite characters and celebrities are at the core of an audience's personal connection with media programs through continuous and simultaneous parasocial interaction, identification, and empathy (Frost, 2010; Macionis & Sparks, 2009). Despite viewers' individual variation in the degree of audience involvement, an empathetic and emotional relationship is gradually developed and maintained over time after repeated exposure to media personalities (Beeton, 2005; Kim & Long, 2012; Riley et al., 1998).

In particular, serialized TV dramas such as *My Love From the Star*, in which personal and domestic narratives are often central to storylines, touch on commonly shared human interests and experiences that anyone can identify with in everyday life (Kim & Long, 2012). Through the viewing process, audiences develop

Table 6. Results of structural equation modeling analysis (N = 430).

Regression path	Standardized coefficient	S.E.	Composite reliability (t-value)	p-value	Decision
H1-1 (V1 → EI)	.27	.07	3.70***	.000	Support
H1-2 (V2 → EI)	.91	.11	8.55***	.000	Support
H1-3 (V3 → EI)	.33	.06	5.42***	.000	Support
H2-1 (V1 → RR)	−.07	.06	−1.25	.211	Reject
H2-2 (V2 → RR)	.84	.11	7.60***	.000	Support
H2-3 (V3 → RR)	.03	.05	.58	.563	Reject
H3-1 (V1→BI)	.48	.07	7.14***	.000	Support
H3-2 (V2 → BI)	−.14	.11	−1.25	.211	Reject
H3-3 (V3 → BI)	.16	.05	3.45	.000	Support
H4-1 (V1 → IT)	−.11	.10	−1.09	.278	Reject
H4-2 (V2 → IT)	−.01	.16	−.05	.962	Reject
H4-3 (V3 → IT)	−.12	.07	−1.72	.086	Reject
H5 (EI → BI)	.29	.07	4.19***	.000	Support
H6 (RR → BI)	.17	.09	1.96*	.049	Support
H7 (EI → RR)	.39	.07	5.68***	.000	Support
H8 (EI → IT)	.09	.12	.74	.461	Reject
H9 (RR → IT)	.68	.11	6.24***	.000	Support
H10 (BI → IT)	.60	.13	4.44***	.000	Support
$\chi^2(386) = 1033.38$ (p = .000); TLI = 0.90; CFI = 0.91; RMSEA = 0.06; NFI = 0.87; GFI = 0.86;AGFI = 0.83					

H: hypothesis; V1: character and celebrity; V2: message; V3: beauty of backdrops; EI: emotional involvement; RR: referential reflection; BI: behavioral involvement; IT: behavioral intention to be a film tourist.
*** $p < 0.001$, ** $p < 0.01$, * $p < 0.05$.

feelings of intimacy and/or a bond with the characters and celebrities as if they were close friends, and thus may feel as if they were participating in a real story even though they are aware that it is fiction. In addition, audiences often apply characters' situations and behavior to the understanding of their own lives and the lives of others, which is considered to be referential reflection and behavioral involvement in this study (Giles, 2002; Rittichainuwat & Rattanaphinanchai, 2015). Similarly, the role of celebrity endorsement has also been discussed in the context of symbolic communications theory and a meaning transfer model in which celebrity icons are interpreted as symbolic meaning-makers (McCracken, 1989) and are thus conveyed to audiences.

This finding can be discussed in the context of parasocial interaction. That is, the interaction between the audience and representations of humans shown in the media, such as celebrities, is developed as a form of parasocial relationship (Horton & Wohl, 1956). Through the parasocial relationship, companionship and personal identity are formed. As a result of this, the audience feels companionship with the actors and identifies with them; this finally leads to dependency on the media (Giles, 2002; Schiappa, Allen, & Gregg, 2007). Thus, the selection of popular and well-matched characters is important to the success of the TV drama and, consequently, the film tourism business.

Second, it is interesting to note that the backdrop of a TV drama (i.e. its filming location) is a significant contributor to emotional involvement only, and there is no statistical confirmation of its effect on referential reflection and behavioral involvement. To some extent this result is unexpected because a study by Kim and Assaker (2014) empirically proved a positive causal relationship between the perceived values of TV dramas and audience involvement, although their study investigated the overall relationship between the named two variables. It is hard to provide an immediate interpretation of this finding and thus draw definitive conclusions from it, given the fact that multiple layers, structures, and symbolic meanings between filmed locations and audiences are manifested through the complex imaginative and affective involvement of audiences (Reijnders, 2010, in press; Roesch, 2009). It may be interpreted in regard to the following: (1) each dimension of audience involvement is interconnected with rather than totally separated from its own territory, and simultaneous interactions among the three dimensions of audience involvement are common when engaging with a media program (Sood, 2002); and (2) this study confirms that emotional involvement is a direct precursor of referential reflection and behavioral involvement. As this study is the first of its kind, more research is needed to further investigate the complex structural and causal relationships between the multiple dimensions of the perceived values of films and audience involvement in the broader context of film tourism.

Third, and related to the previous point, emotional involvement significantly affected referential reflection and behavioral involvement in the context of film tourism, as suggested by previous media audience studies (Sood, 2002; Sood & Rogers, 2000). This finding is particularly important for future film tourism studies as it was the first attempt to prove the influence of emotional involvement on the two other dimensions of audience involvement. This re-confirms that an audience's emotional engagement and identification with the values of TV dramas plays a crucial role in enhancing audience viewing experiences that lead to the development of referential reflection and behavioral involvement (Carl et al., 2007; Kim & Assaker, 2014; Macionis & Sparks, 2009).

Fourth, referential reflection and behavioral involvement showed a significant direct effect on behavioral intention. This result confirms the findings of some previous studies suggesting that the higher the level of involvement with a media program, the stronger the behavioral intention to visit film tourism locations (Kim, 2012a; Lee et al., 2008; Reijnders, 2010; Roesch, 2009; Wong & Lai, 2015; Yen & Teng, 2015). It is noteworthy, however, that emotional involvement had no statistically significant effect on behavioral intention. As no previous studies have statistically tested which dimensions of audience involvement directly influence an audience's behavioral intention to visit film tourism locations, there is no single and simple explanation for this finding. However, one interpretation posits that unless audiences develop referential reflection and behavioral involvement through emotional involvement with a media program, it is unlikely that they will decide to visit locations featured in the media program in question. This finding is the main contribution of this paper, as previous studies simply highlight the significance of emotional engagement in film tourism, with little investigation of the structural complexity of audience involvement in the context of film tourism (Lee et al., 2008; Wong & Lai, 2015; Yen & Croy, in press; Yen & Teng, 2015).

Additionally, even though behavioral involvement was highly correlated to behavioral intention (t-value = 4.44, $p < .001$), they are not the same construct; that is, it was found that behavioral involvement can be achieved through excellent production values and the level of emotional involvement and centrality. However, the effect of the production value on predicting behavioral intention was different. Therefore, the two constructs do not overlap. This indicates that future studies should include both constructs in exploring visual media consumption models.

Finally, all three perceived values did not directly influence behavioral intention, even though most paths showed indirect effects, through other paths, to behavioral intention. However, previous studies have confirmed that the significant values and meanings of a TV drama's attributes, such as characters and celebrities, and messages and backdrops, become symbolically meaningful icons only through the audience's personalized emotional attachment to them during and after previous viewing experiences of the program (Couldry, 1998; Reijnders, 2010; Wong & Lai, 2015).

In reality, audiences exposed to TV dramas do not immediately decide to travel to the film destinations. Instead, they go through an elaborate likelihood process (Cacioppo & Petty, 1984), involvement (Kim & Assaker, 2014; Zaichkowsky, 1984), or commitment (Cheng, Hung, & Chen, 2016), to interpret the meanings of the visual media and then to personalize the meanings. In addition to this, actual overseas travel requires an overcoming of constraints, such as structural factors (e.g. crowding, safety, travel cost, no time), interpersonal factors (e.g. people to travel with), and situational factors (e.g. weather, poor health, scheduling conflicts). This is explained by the phenomenon of parasocial interaction, which describes the relationship between media and media figures. Media researchers (Giles, 2002; Schiappa et al., 2008) assert that audiences become friendlier and identify with characters through consistent interaction with a program. As a consequence of this, it takes some time to decide to actually travel overseas to film destinations.

Academic and practical implications

Despite previous contributions to film tourism literature, the need for theoretical and conceptual contributions has been strongly suggested (Connell, 2012; Reijnders, in press). In addition, scholarly efforts to test a complex conceptual model covering the diverse key constructs of film tourism and their causal relationships still remain limited. Thus, the academic merit of this study has been to conceptually propose and empirically test a TV drama consumption model of film tourism in the context of the mainland Chinese audience's viewing experience with one of the most successful recent K-dramas, namely *My Love from the Star*. Future film tourism studies can beneficially incorporate the key findings and contributions of this study in studies of other countries or in different research contexts, because the production content of each media program can vary with the production country's cultural background and production style (Hudson, Wang, & Gil, 2011; Kim & O'Connor, 2011; Su et al., 2011). In addition, further modifications of the current TV drama consumption model of film tourism, reflecting other key concepts such as cultural proximity, authenticity, transformation, and nostalgia and memory, will be welcome.

This study has several important practical implications. First, it indicates that TV dramas may not lead to tourism or economic consumption in regard to a film destination because three dimensions of the perceived values of TV dramas did not significantly influence behavioral intention to participate in film tourism. Considering that the duration of broadcast of TV dramas can be a few months, a certain amount of diffusion time is needed to allow the TV dramas to penetrate new countries or cultural territories through diverse promotional channels, including word-of-mouth and the Internet (Zufryden, 2000). Film destination marketers need to establish appropriate marketing and branding strategies for potential film tourism destinations prior to, during, and after the production of media programs, because there is no guarantee that audiences who are exposed to media programs will visit film tourism destinations. Although film tourism is a global tourism phenomenon with a high demand, a host film destination is not able to satisfy this demand without an overall plan to accommodate tourists at different points in time.

The main finding of this study shows that the effects of dramatized characters (actors) and celebrities on subsequent consumption items were significant. Audiences watching TV dramas may be more interested in the popularity of actors than in the storyline or filming locations. This suggests that it is more effective to utilize actors than scenery or storyline in promoting a film tourism destination; advertisements including photos of actors and characters will be most effective. In addition, actor-guided tours will generate interest and attention. Further, the effects of celebrity on a tourist can be maximized when a good match is achieved between the characteristics of the endorser and the tourist's self-image (Lee et al., 2008; Magnini, Honeycutt, & Cross, 2008; Yen & Croy, in press; Yen & Teng, 2015).

As the findings show, TV drama values did not influence behavioral intention, even though they were influential during mediating stages, such as involvement. The results indicate that TV drama audiences do not immediately become involved with a TV drama but are steadily committed through assimilation or a deep emotional involvement process including identification and referential reflection. Therefore, film destination marketers need to promote the film tourism destinations until the audience is assimilated and emotionally involved, and should develop film tourism products which can satisfy film tourists' needs and expectations. For example, with assumptions of a TV drama's success, there is a need for proactive effort, such as the preservation of film sets, the creation and updating of an Internet website, the posting up of film locations on the website, speedy replies to fans, and the introduction of a package tour of film locations. As actual film tourists may be only a small part of the audience, film destination marketers must make consistent communication efforts.

Study limitations and suggestions for future studies

This study is subject to some limitations. The survey was conducted in the summer of 2014, after *My Love From the Star* swept the Asian region. As the popularity of the drama was established and the actors were known, the importance of celebrity might be exaggerated compared with other perceived values of the program. This study highlighted the role of TV drama consumption value in predicting consequent outcomes, such as future intention. However, this study did not measure specific attitudes, such as motivations, constraints, or influential factors in decision making, which can determine the actual levels of film tourism that occur. Thus, future studies need to include those variables in order to allow for an understanding of the role of film tourists' attitudes and behaviors in regard to film travel.

Finally, current visual media consumption models do not find any significant difference between TV drama values and behavioral intention to actually participate in film tourism. It was discussed that there is a need for immersion time in this respect. Therefore, a longitudinal study is required to investigate whether or not the relationships between constructs are significant when popularity prevails more widely after the end of broadcast by the visual media. Since most film tourism studies have used cross-sectional study samples, there is a need to attempt longitudinal research in order to explore the effects of visual media on tourists' visits to film destinations.

Disclosure statement

No potential conflict of interest was reported by the authors.

Funding

This work was fully funded by The Hong Kong Polytechnic University (Project No: PolyU 1-ZE3J).

References

Babbie, E. (1995). *The practice of social research* (7th ed.). Belmont, CA: Wadsworth Publishing company.

Bae, H., & Lee, B. (2004). Audience involvement and its antecedents: An analysis of the electronic bulletin board messages about an entertainment-education drama on divorce in Korea. *Asian Journal of Communication*, *14*(1), 6–21. doi:10.1080/0129298042000195125

Bagozzi, R. P., & Yi, Y. (1991). Multitrait-multimethod matrices in consumer research. *Journal of Consumer Research*, *17*, 426–439. doi:10.1086/jcr.1991.17.issue-4

Beeton, S. (2005). *Film-induced tourism*. Clevedon: Channel View Publications.

Buchmann, A., Moore, K., & Fisher, D. (2010). Experiencing film tourism: Authenticity and fellowship. *Annals of Tourism Research*, *37*(1), 229–248. doi:10.1016/j.annals.2009.09.005

Cacioppo, J. T., & Petty, R. E. (1984). The elaboration likelihood model of persuasion. *Advances in Consumer Research*, *11*, 673–675.

Carl, D., Kindon, S., & Smith, K. (2007). Tourists' experience of film locations: New Zealand as 'MiddleEarth'. *Tourism Geographies*, *9*(1), 49–63. doi:10.1080/14616680601092881

Cheng, T.-M., Hung, S.-H., & Chen, M.-T. (2016). The influence of leisure involvement on flow experience during hiking activity: Using psychological commitment as a mediate variable. *Asia Pacific Journal of Tourism Research*, *21*(1), 1–19. doi:10.1080/10941665.2014.1002507

Chosun Ilbo. (2016, March 28). Like main actors. Retrieved April 1, 2016, from http://news.chosun.com/site/data/html_dir/2016/03/28/2016032803114.html

Comrey, A., & Lee, H. (1992). *A first course in factor analysis* (2nd ed.). Hillsdale, NJ: Erlbaum.

Connell, J. (2012). Film tourism – Evolution, progress and prospects. *Tourism Management*, *33*(5), 1007–1029. doi:10.1016/j.tourman.2012.02.008

Fornell, C., & Larcker, D. (1981). Evaluating structural equation models with unobservable variables and measurement errors. *Journal of Marketing Research*, *18*, 39–50. doi:10.2307/3151312

Frost, W. (2010). Life changing experiences: Film and tourists in the Australian outback. *Annals of Tourism Research*, *37*(3), 707–726. doi:10.1016/j.annals.2010.01.001

Giles, D. (2002). Parasocial interaction: A review of the literature and a model for future research. *Media Psychology*, *4*, 279–305. doi:10.1207/S1532785XMEP0403_04

Grönlund, K., & Strandberg, K. (2014). Online panels and validity: Representativeness and attrition in the Finnish eOpinion panel. In M. Callegaro, R. Baker, J. Bethlehem, A. S. Göritz, J. A. Krosnick, & P. J. Lavrakas (Eds.), *Online panel research: A data quality perspective* (pp. 86–103). New York, NY: Wiley.

Hair, J., Anderson, R., Tatham, R., & Black, W. (2009). *Multivariate data analysis* (7th ed.). Upper Saddle River, NJ: Prentice Hall.

Horton, D., & Wohl, R. R. (1956). Mass communication and para-social interaction. *Psychiatry*, *19*, 215–229.

Hudson, S., Wang, Y., & Gil, G. M. (2011). The influence of a film on destination image and the desire to travel: A cross-cultural comparison. *International Journal of Tourism Research*, *13*(2), 177–190.

Kim, S. (2012a). Audience involvement and film tourism experiences: Emotional places, emotional experiences. *Tourism Management*, *33*(2), 387–396.

Kim, S. (2012b). The impact of TV drama attributes on touristic experiences at film tourism destinations. *Tourism Analysis*, *17*(5), 573–585. doi:10.3727/108354212X1348 5873913804

Kim, S., & Assaker, G. (2014). An empirical examination of the antecedents of film tourism experience: A structural model approach. *Journal of Travel & Tourism Marketing*, *31*(2), 251–268. doi:10.1080/10548408.2014.873316

Kim, S., & Long, P. (2012). Touring TV soap operas: Genre in film tourism research. *Tourist Studies*, *12*(2), 173–185. doi:10.1177/1468797612449249

Kim, S., & O'Connor, N. (2011). Film tourism locations and experiences: A popular Korean television drama production perspective. *Tourism Review International*, *15*(3), 243–252. doi:10.3727/154427211X13216636845701

Kim, S. S., Agrusa, J., Lee, H., & Chon, K. (2007). Effects of Korean television dramas on the flow of Japanese tourists. *Tourism Management*, *28*(5), 1340–1353. doi:10.1016/j.tourman.2007.01.005

Lee, S., Scott, D., & Kim, H. (2008). Celebrity fan involvement and destination perceptions. *Annals of Tourism Research*, *35*(3), 809–832. doi:10.1016/j.annals.2008.06.003

Lee, S. H. (2015, September 7). Korean food is going to China as it is. *KyunghyangBiz*. Retrieved April 1, 2016, from http://biz.khan.co.kr/khan_art_view.html?artid=201509072216435&code=920501& med=khan

Macionis, N., & Sparks, B. (2009). Film-induced tourism: An incidental experience. *Tourism Review International*, *13*(2), 93–101. doi:10.3727/154427209789604598

Magnini, V., Honeycutt, E., & Cross, A. (2008). Understanding the use of celebrity endorsers for hospitality firms. *Journal of Vacation Marketing*, *14*(1), 57–69. doi:10.1177/1356766707084219

McCracken, G. (1989). Who is the celebrity endorser? Cultural foundations of the endorsement process. *Journal of Consumer Research*, *16*(3), 310–321. doi:10.1086/jcr.1989.16.issue-3

Nunnally, J. (1978). *Psychometric theory* (2nd ed.). New York, NY: McGraw Hill.

Papa, M. J., Singhal, A., Law, S., Pant, S., Sood, S., Rogers, E. M., & Shefner-Rogers, C. L. (2000). Entertainment-education and social change: An analysis of parasocial interaction, social learning, collective efficacy, and paradoxical communication. *Journal of Communication*, *50*(4), 31–55. doi:10.1111/jcom.2000.50.issue-4

Podsakoff, P. M., & MacKenzie, S. B. (2003). A critical review of construct indicators and measurement model misspecification in marketing and consumer research. *Journal of Consumer Research*, *30*(September), 199–218. doi:10.1086/376806

Reijnders, S. (2010). Places of the imagination: An ethnography of the TV detective tour. *Cultural Geographies*, *17*(1), 37–52. doi:10.1177/1474474009349998

Reijnders, S. (in press). Stories that move: Fiction, imagination, tourism. *European Journal of Cultural Studies*.

Riley, R., Baker, D., & Van Doren, C. (1998). Movie-induced tourism. *Annals of Tourism Research*, *25*(4), 919–935. doi:10.1016/S0160-7383(98)00045-0

Rittichainuwat, B., & Rattanaphinanchai, S. (2015). Applying a mixed method of quantitative and qualitative design in explaining the travel motivation of film tourists in visiting a film-shooting destination. *Tourism Management*, *46*, 136–147. doi:10.1016/j.tourman.2014.06.005

Roesch, S. (2009). *The experiences of film location tourists*. Bristol: Channel View Publications.

Schiappa, E., Allen, M., & Gregg, P. B. (2007). Parasocial relationships and television: A meta-analysis of the effects. In R. W. Preiss, B. M. Gayle, N. Burrell, M. Allen, & J. Bryant (Eds.),

Mass media effects research: Advances through meta-analysis (pp. 301–314). New York: Routledge.
Sood, S. (2002). Audience involvement and entertainment-education. *Communication Theory*, *12*(2), 153–172.
Sood, S., & Rogers, E. (2000). Dimensions of parasocial interaction by letter-writers to a popular entertainment-education soap opera in India. *Journal of Broadcasting & Electronic Media*, *44*(3), 386–414. doi:10.1207/s15506878jobem4403_4
Su, H. J., Huang, Y., Brodowsky, G., & Kim, H. J. (2011). The impact of product placement on TV-induced tourism: Korean TV dramas and Taiwanese viewers. *Tourism Management*, *32*(4), 805–814. doi:10.1016/j.tourman.2010.06.020
Wong, J. Y., & Lai, T. C. (2015). Celebrity attachment and behavioural intentions: The mediating role of place attachment. *International Journal of Tourism Research*, *17*(2), 161–170. doi:10.1002/jtr.1974
Yen, C. H., & Croy, G. W. (in press). Film tourism: Celebrity involvement, celebrity worship, and destination image. *Current Issues in Tourism*. doi:10.1080/13683500.2013.816270
Yen, C. H., & Teng, H. Y. (2015). Celebrity involvement, perceived value, and behavioural intentions in popular media-induced tourism. *Journal of Hospitality & Tourism Research*, *39*(2), 225–244. doi:10.1177/1096348012471382
Zaichkowsky, J. L. (1984). Measuring the involvement construct. *Journal of Consumer Research*, *12*(3), 341–352. doi:10.1086/208520
Zhang, S. Y., & Choi, M. Y. (2015). A study on Chinese consumers perception changing for Korean cosmetics brand before and after watching Hallyu drama. *Journal of Brand Design Association of Korea*, *13*(3), 139–148. doi:10.18852/bdak.2015.13.3.139
Zufryden, F. (2000). New film website promotion and box office performance. *Journal of Advertising Research*, *40*(1–2), 55–64. doi:10.2501/JAR-40-1-2-55-64

Segmentation of potential film tourists by film nostalgia and preferred film tourism program

Seongseop (Sam) Kim and Sangkyun (Sean) Kim

ABSTRACT

The current study sought to identify the role of nostalgia in film tourism, specifically in relation to the development of preferred film tourism products and activities for potential Korean film tourists visiting the shooting locations of Hong Kong films made from the 1970s to the late 1990s. The findings suggest that the concept of nostalgia in the context of film tourism is multidimensional and consists of five identified domains. Of these domains, "reminiscence of mimicking", "memory of film backdrops", and "memory of Hong Kong history and culture" were highly significant in explaining familiarity with film destinations and the intention to engage in nostalgia film tourism. Differences in characteristics of respondents' profiles were categorized into clusters generated using extracted domains. Among the four extracted clusters, the high film nostalgia group (Cluster 2) and the low film nostalgia group (Cluster 3) showed completely different characteristics. These differences included preferred activities or tour programs and socio-demographic profiles. As a result, this study found film nostalgia to be an important segmentation variable. The results of this study can be prudently used by film destination marketers to develop new products or programs and to satisfy their on-site experiences.

Introduction

Hong Kong films enjoyed huge popularity throughout Asia before the 2000s, and Hong Kong became the center of the Asian film industry, known as 'Hollywood in Asia' (Hsu & Song, 2013; Weng, 2008). In those days, watching a Hong Kong film in a theater was considered a great leisure and entertainment activity (H. Lee, 2006). Asian film audiences grew up watching Hong Kong action movies featuring martial arts, gangsters, and crime stories and, less commonly, romance or comedy movies (Pan & Ryan, 2013). For example, Bruce Lee became an iconic star of Hong Kong films in his collection of martial arts films, including *The Big Boss* (1971), *Fist of Fury* (1972), *The Way of the Dragon* (1972), *Enter the Dragon* (1973), and *Game of Death* (1978). Lee was a legend in Asia, especially among young male audiences, who mimicked his gestures, voice, and clothing. Many of these audience members have maintained strong memories of Lee's films, which they saw at a time when other entertainment was limited (Chosun Ilbo, 2016).

The popularity of Hong Kong films in Korea at that time led to a kind of social syndrome (Chosun Ilbo, 2016) as Korean viewers exposed to Hong Kong movies often mimicked the actions of Hong Kong actors and actresses by performing martial arts, wearing sunglasses, smoking cigarettes, posing like gangsters, wearing the same make-up, and often imitating scenes word-for-word (H. Lee, 2006; Kim, 2015). These audience members are now in their forties or older. In particular, Bruce Lee's films have become a trigger for their memories and nostalgia for a cherished past. As a result of their memories of Lee's films, middle-aged people often visit the Hong Kong Heritage Museum, which exhibits Lee's memorabilia. Lee's iconic statue on the "Avenue of Stars" is a popular tourist attraction, where middle-aged nostalgia-driven tourists go to reminisce about his films and re-experience those poor but happy days by re-enacting Lee's actions and even taking photographs of their re-enactments.

As time goes by, the films that an individual views are stored in his or her memory. Nostalgic sentiments resurface whenever the stories, songs from the original soundtracks, actors, or props come to mind. Memories of a film location can be a crucial guide in the selection of a tourism destination given that films that create nostalgia form positive images of such locations in the minds of audience members (Alderman, Benjamin, & Schneider, 2012; Kim, 2012a, 2012b; Kim & Assaker,

2014; Liou, 2010; Reijinders, in press; Robinson, 2015; Suni & Komppula, 2012). Despite the significance of nostalgia in such a context, few studies have explored the potential of preferred film tourism products and activities that must be tailored to meet this specific need, that is, nostalgia. It is also noteworthy that although the Hong Kong government has developed film destinations such as the "Avenue of Stars" and a restoration of Bruce Lee's old residence, these efforts have been made without research examining the role of nostalgia in film tourists' preferences for products and activities related to those destinations.

This study has two objectives as it attempts to investigate the role of nostalgia in film tourism. The first is to assess whether film nostalgia has similar influences on the film destination, the intention to engage in film nostalgia tourism by traveling to the film destination, and the influence of previous films on the decision to travel to the destination. The second objective is to attempt to use film nostalgia as an indicator to segment those who were exposed to these films and to identify which variables are meaningful for a detailed explanation of film nostalgia tourism clusters. The variables used include involvement, image, familiarity, perceived cultural proximity, preferred activities at the film nostalgia destination, preferred tour programs or products to induce film nostalgia tourism, sociodemographic characteristics, film fandom-related characteristics, and preferred film nostalgia tourism characteristics at the film destination. These efforts help to develop preferred film tourism products and programs targeted at the film audience.

Film, nostalgia, and film tourism

Nostalgia has been recognized as an important motivational factor for tourism in general (see for example Hsu, Cai, & Wong, 2007; Hunt & Johns, 2013; Leong, Yeh, & Chang, 2015a; Leong, Yeh, Hsiao, & Huan, 2015b; Sellicka, 2004; Wang, Li, & Chou, 2014; Yeh, Chen, & Yeh, 2012). It has also been described as an important factor in understanding the film tourism phenomenon (see for example Alderman et al., 2012; Bandyopadhyay, 2008; Kim, 2012a; Kim & Assaker, 2014; Liou, 2010; Reijinders, in press; Rittichainuwat & Rattanaphinanchai, 2015; Robinson, 2015; Suni & Komppula, 2012; Vesey & Dimanche, 2003). However, most film tourism studies have been limited to explanations of nostalgic sentiments or activities in a film tourism destination.

Films can stimulate nostalgia when they tell stories of the "good old days" or when they are analogous to one's own story or that of one's community. The content and storylines of old films stored in an individual's memory can often serve as nostalgic vestiges of times past and can encourage the viewer to form an emotional connection with stories or scenes. In the context of film tourism, nostalgia is often used as both a marketing tool and an image enhancer for destinations (Robinson, 2015). Film sets or locations used to shoot film or television dramas become nostalgia-stimulating spaces for audiences, whose memories are evoked by the stories and events associated with the shooting locations (Roesch, 2009; Wray & Croy, 2015)

Films have underlying functions that are relevant to nostalgia. First, the scenes we watch can make us reminisce about the past, and our memories take on positive or negative affective signatures. For example, films such as *Saving Private Ryan* (1998), *Life is Beautiful* (1997), *The Tin Drum* (1979), and *Band of Brothers* (2001), all of which depict former wars, can spur seniors with relatable real-world experiences to recollect traumatic past events (Kessous, 2014). Second, films can represent the popular trends or fads of an era. Some films screened in the 1970s and 1980s were war-themed movies with storylines that unfolded in the contexts of the Vietnam, Korean, and Cold Wars. Therefore, the films evoke nostalgia of the era's societal ethos or life at that time, especially in those who lived in those days.

The personal nostalgic sentiment acts as a metaphor of creating and thus perceiving a place attachment with a shooting location while watching film or television dramas (Kim, 2012a) or the so-called 'topophilia', referring to all emotional connections between physical environment (e.g. film shooting location) and human being (Tuan, 1980). As a result, it leads to a positive bond between audiences and familiarity with the film locations (Croy & Heitmann, 2011; Kim, Kim, & Heo, 2015; Pan & Ryan, 2013; Rittichainuwat & Rattanaphinanchai, 2015). The psychological attachment may stimulate some audiences to become film tourists who actually visit film locations.

Indeed, nostalgia can attract media audiences to a country in which a film was shot or set. For example, during the 1980s, Australian films, to attract American attention, portrayed substantial footage of Australia's natural attributes and interesting storyline: *Crocodile Dundee* in Kakadu National Park illustrates this point well. This example of interesting storyline content and typical environmental landscapes in Australian films seems to foster a nostalgic desire for 'paradise lost' in film tourists (Riley & Van Doren, 1992). In this regard, although there is no single and simple explanation of the above, psychological motivational factors beyond the natural beauty of these shooting locations in Australia would be understood as a form of nostalgia or a sense of loss. Essentially, films create nostalgia, which can induce film nostalgia tourism.

The conceptual framework of this research is presented in Figure 1. Korean audiences who were exposed to Hong Kong films from the 1970s to the late 1990s are supposed to have those films deeply rooted in their memories for a number of specific reasons. First, from the early 1970s to the 1990s, there were few influential media offering entertainment in Korea. Second, Korea's political status, namely that of being under a military regime, meant that people envied democratic societies. Finally, since Korean film-making technology was underdeveloped compared with Hong Kong's, audiences were very enthusiastic about watching Hong Kong films. There were few who did not watch martial arts or gangster films and many people would talk about these films or boast about having seen them (F. L. Lee, 2006; Kim, 2015). Thus, it was a special event during this period to watch Hong Kong films in a movie theatre.

If the films were influential for Korean audiences, these audiences must be imbued with film nostalgia. As a consequence, they have been familiar with and close to film destinations since their exposure to the films. The audiences are likely to experience nostalgic stimuli and intend to visit the film nostalgia destinations. Thus this study attempts to identify the effectiveness of film nostalgia in predicting familiarity with the film destinations and the intention to participate in film nostalgia tourism using multiple regression analysis. Hypothesis (H) 1 was developed to verify the relationships of these variables.

H1: Film nostalgia domains explain familiarity with the film destination, intention to participate in a nostalgia film tourism tour to the film destination, and the influence of nostalgia films on deciding to travel to the film destination.

Film nostalgia is likely to have different dimensions in the continuum of each dimension because the audience members are stimulated by a different set of film nostalgic emotions. Some studies empirically identified the role of audience involvement in the construction of on-site film tourist experiences using structural modeling (Kim, 2012a; Kim & Assaker, 2014). That is, responses to the film's production values (story and character, and location and visual effects) have a direct and positive influence on audience involvement (behavioral involvement, emotional involvement, and referential reflection). Thus, reaction to film nostalgia have a positive correlation with psychological and behavioral involvement with the film.

Most film tourism studies show a consensus that the audience's exposure to a film influences their perception of the image of the filmed location (see for example Connell & Meyer, 2009; Kim, Agrusa, Lee, & Chon, 2007; Redondo, 2012; Riley & Van Doren, 1992) and familiarity with it (see for example Croy, 2010; Pan & Ryan, 2013; Rittichainuwat & Rattanaphinanchai, 2015; Yen & Croy, in press). The reactions to a film can differ according to perceived cultural proximity because the popularity of foreign films can differ by national or cultural background (Huang, 2011; Su, Huang, Brodowsky, & Kim, 2011). For example, some studies (Huang, 2011; Su et al., 2011) examined the role of perceived cultural proximity in determining the appeal of filmed locations of Korean television dramas among Taiwanese audiences, and confirmed that an individual audience's perceived cultural proximity to a country portrayed in a film affected their evaluation of the characters and landscapes associated with the filmed locations.

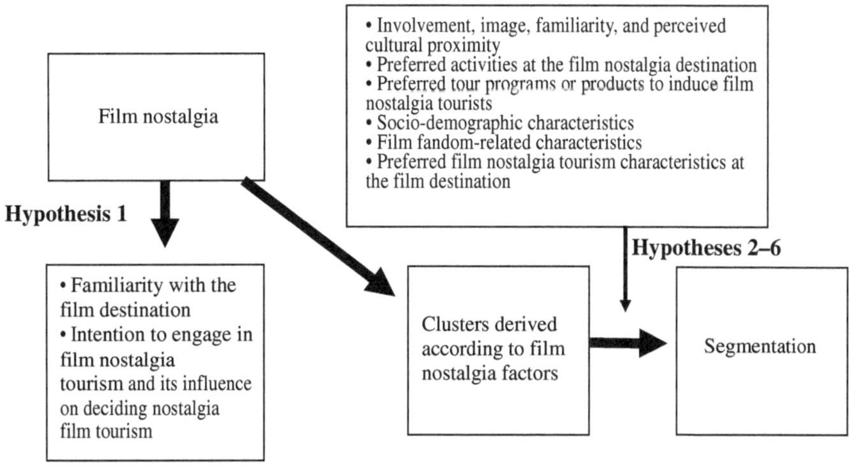

Figure 1. Conceptual framework.

Also, film tourists demonstrate different preferences for activities or tour programs and for products to purchase at the film tourism destination because their attitudinal states vary after watching a film. For example, those who have a high level of film nostalgia and are considered niche "special film tourists", tend to have a higher level of expectation regarding provided services or programs than simply experiencing the locations depicted on the screen or merely gazing at the sights (Buchmann, Moore, & Fisher, 2010; Kim, 2012b; Roesch, 2009). It is expected that these tourists, albeit only some of the audiences, have an eager desire to experience the perspectives of the characters and every prop displayed at the filmed locations (Buchmann et al., 2010; Carl, Kindon, & Smith, 2007; Croy & Heitmann, 2011; Mercille, 2005; Reijnders, 2009).

Film nostalgia tourists show a high level of expectation regarding nostalgic and symbolic experiences at the filmed locations because they have been psychologically and emotionally involved with the plot, the actors, backdrops, songs, and scenic backgrounds of the film (Buchmann et al., 2010; Couldry, 1998; Pan & Ryan, 2013; Roesch, 2009). Thus they become more sentimental by touching props, imitating the motions of actors, reciting lyrics, singing songs from the original soundtrack at the film destination, and waiting for film celebrities at fan meetings, who are categorized as specific or purposeful film tourists who actively seek out personalized and symbolized meanings and values of the filmed locations (Kim, 2012a; Macionis & Sparks, 2009; Rittichainuwat & Rattanaphinanchai, 2015). This emotional connection is linked to an interest in a "sense of place" (Kim, 2012b; Roesch, 2009) or the identity of the film locations (Carl et al., 2007; Couldry, 1998; Schofield, 1996).

Their actions in film locations vary: behaving like causal visitors (Croy & Heitmann, 2011), making a pilgrimage (Bandyopadhyay, 2008; Couldry, 1998; Riley & Van Doren, 1992), experiencing historical reality (Frost, 2006; Kim, 2012a), seeking authenticity (Buchmann et al., 2010; Couldry, 1998; Lee, 2012; Macionis, 2004; Su et al., 2011), and tracking personal reminiscences (Bandyopadhyay, 2008; Macionis & Sparks, 2009; Reijnders, 2016; Robinson, 2015).

They also wish to seek novelty and/or fantasy in visiting a location portrayed in a film (Beeton, 2010; Liou, 2010; Macionis, 2004; Macionis & Sparks, 2009; Riley & Van Doren, 1992). They are also interested in seeking self-expression to disclose interests that differ from those of others, and prestige to earn recognition (Macionis & Sparks, 2009). Some film nostalgia tourists are motivated to see spectacular natural scenery or landscapes (Connell & Meyer, 2009; Riley, Baker, & Van Doren, 1998; Soliman, 2011; Tooke & Baker, 1996).

Film destination businesses therefore take advantage of film nostalgia to attract tourists to the film destinations. For example, the eight films in the *Harry Potter* series (2001–2011) have successfully created intense nostalgia in their fans, especially with child audiences (Lee, 2012). Fourteen years after the release of the first film, these audiences, whose members are no longer children, have nostalgic memories of the films' backdrops, props, characters, special effects, and stories. Their sensational reminiscences perhaps motivate them to visit a number of authentic indoor and outdoor sets associated with the *Harry Potter* series that exhibit an array of arts and graphics, costumes, special effects, hand-crafted props, the "Hogwarts Express", soundstages, and backlots. Some hotels (e.g. The Georgian House Hotel in London, United Kingdom) have introduced *Harry Potter*-themed "Wizard Chambers" .

As a result, film nostalgia tourists can be spotlighted as members of a newly surfacing niche market who seek sensational and personalized experiences based on their own interpretation, novelty, fantasy, gratification, imagination, pleasure, romance, nostalgia, and memories associated with the screen (Frost, 2006; Kim, 2012a; Macionis & Sparks, 2009; Pretes, 1995). The nostalgic motivation to travel to a film backdrop results in expectations about the attributes of the actual setting. Thus, while visiting film tourism destinations they expect to experience in person what they vicariously felt in front of the screen, which leads to different expectations of the quality of authentic experience, service quality, preferences, and satisfaction at the film destinations. This finding implies that film destination marketers should understand the characteristics of sociodemographics, attitude, behavior, previous involvement, influence of nostalgic films, and preferences for products or tour programs.

As Figure 1 shows, it is expected that a group of film nostalgia tourists can be divided into clusters according to their film nostalgia domains, which are extracted by factor analysis. Segmentation analysis should be conducted before destination marketing strategies are developed, which makes it feasible to choose a target market by understanding the characteristics of each segmented market. The most commonly used variables in segmentation analyses are sociodemographic indicators such as income, nationality, gender, cultural background (see for example Kim & Agrusa, 2008), psychological variables behavioral variables such as the amount of usage and economic variables such as expenditure (see for example Becken & Gnoth, 2004).

This study selected film nostalgia as an indicator of segmentation because, as described above, it is of value for the study to identify different characteristics of the derived segments. This study assumes that clusters extracted as a result of cluster analysis may demonstrate different characteristics of derived nostalgia domains. Few studies, if any, have used film nostalgia as a segmentation variable. Furthermore, this study was carried out to identify whether the clusters show differences in such variables as involvement, perceived image, familiarity, perceived cultural proximity, preferred activities at the film nostalgia destination, preferred tour programs or products to induce film nostalgia tourism, sociodemographic characteristics, and film fandom-related characteristics. The outcomes of the analyses will be used to investigate the features of film nostalgia clusters and establish appropriate marketing strategies for film nostalgia tourism. Therefore, H2–H6 were developed as follows.

H2: Each film nostalgia cluster has distinctive characteristics of involvement, image, familiarity, and perceived cultural proximity relative to those of other clusters.

H3: Each film nostalgia cluster has distinctive characteristics of preferred activities at the film nostalgia destination relative to those of other clusters.

H4: Each film nostalgia cluster has distinctive characteristics of preferred film tourism programs or products to induce film nostalgia tourists relative to those of other clusters.

H5: Each film nostalgia cluster has distinctive sociodemographic and film fandom-related characteristics relative to those of other clusters.

H6: Each film nostalgia cluster has distinctive characteristics of preferred film nostalgia tourism relative to those of other clusters.

Research methods

In order to develop items pertaining to nostalgia, specifically about Hong Kong films, an extensive review of journals, magazines, and books relevant to critiques of films, social culture and film, social trends and motives, and the characteristics of films between the 1970s and the late 1990s was conducted (see for example F.L. Lee, 2006; H. Lee, 2006; Pan & Ryan, 2013; Robinson, 2015; Suni & Komppula, 2012; Wong & McDonogh, 2001). On the basis of this literature review, a preliminary questionnaire was developed. In-depth interviews were then conducted to identify whether the items were consistent with the results of the literature review. Twenty Korean residents who reported enjoying Hong Kong movies made from the 1970s to the late 1990s participated in the in-depth interviews. Given the significance of audience involvement in the context of film tourism suggested by Kim (2012a), the topics covered included the respondents' emotional and behavioral involvement, the effects of the films on the respondents and their lives, the social motifs, the circumstances of the domestic film industry, the respondents' intention to engage in film tourism in Hong Kong, and their preferred activities and film tourism programs.

Based on the above, the items chosen to measure nostalgia after watching Hong Kong films included remembering scenes or backdrops; images of Hong Kong featured in the films; the films' stories, feelings, music, and background images; the visual effects of kung fu films (e.g. flying in a kung fu movie); mimicking or gossiping after watching films; and interest in Hong Kong's culture or history. To develop the film nostalgia items previous studies in consumer behavior were reviewed (Holak, 2013; Holbrook, 1993; Marchegiani & Phau, 2013).

The items chosen to measure preferred tourism products for attracting nostalgia-driven film tourists included film destination tour programs, visual or mobile programs, and special event programs. The items chosen to measure preferred tourism activities in the nostalgia-driven film tourism program included visiting film sites, fan meeting events, and tasting Cantonese food (e.g. visiting the restaurants or cafés featured in the Hong Kong films). The items chosen to measure perceived cultural proximity, after revising those used in previous studies, included similarity of tradition, organizational culture, customs, food, and film stories between Hong Kong and Korea (see for example Burch, 2002; F.L. Lee, 2006; Huang, 2011; Su et al., 2011).

Questions pertaining to fandom included level of fandom for Hong Kong films in the 1970s through to the late 1990s, and level of current fandom for Hong Kong films. The items used to describe familiarity with Hong Kong from watching Hong Kong films measured an increase in the respondents' sense of closeness or friendliness toward Hong Kong after watching Hong Kong films, as adapted from previous studies (Chen & Lin, 2012; Horng, Liu, Chou, & Tsai, 2012). Six items of perceived cultural proximity indicating similarity of culture were used. The items mentioned above (with two exceptions mentioned below) were measured on a 5-point Likert scale (1 = strongly disagree; 5 = strongly agree). Gender, age, and occupation were measured as categorical data, while the number of films that a respondent watched (from a list of 30) and the number of movie stars that a respondent could recognize (from a list of 11) were measured as "yes" (recognize) or "no" (do not recognize) questions.

The questions were then revised using pilot testing among 50 Korean residents who watched Hong Kong films in the past days. Data were collected from October 1–15, 2015, using an online company with 1.12 million panelists. Based on their comments on the questionnaires, a small modification was made to some nostalgia items. A main survey was then conducted using an online survey tool chosen for its advantages, such as rapid collection of samples from a targeted population, the ability to control sample selection using planned quotas, and the user-friendliness of the website (Grönlund & Strandberg, 2014). The restrictions for the selection of a targeted sample included individuals who were 40 years of age or older, had not visited Hong Kong, had watched at least two of a list of 30 Hong Kong films released between the 1970s and the late 1990s, and recognized at least one of a list of 11 film stars from that time. The respondents were exposed to Hong Kong films dating from the 1970s to the late 1990s. As mentioned in the introduction and the literature review, because watching Hong Kong films had become a societal syndrome during those days in Korea, the sample group did not represent the impossible rationales of that group's interest in pursuing film tourism in Hong Kong to seek their memories of the past days.

Of the 620 questionnaires completed in the survey, 610 were used for further data analyses; the remaining 10 had multiple missing values. For the data analyses, descriptive analysis was conducted for demographic features, the number of Hong Kong films released from the 1970s to the late 1990s, and the famous stars at the time.

Results

Respondent profiles

A frequency analysis of the respondents' demographic profiles showed that 51.3% were male, 84.4% were married, and 30.6% were in their early fifties. With regard to occupation, 45.4% were company workers and 22.3% were housewives. Regarding residence area, 62.8% lived in a metropolitan area or large city and 32.3% lived in a medium-size or small city.

In terms of their experiences viewing famous Hong Kong films released between the 1970s and the late 1990s, more than half of the respondents reported having watched 10 of the 30 films. As a similar question to ask exposure to these films, in terms of the number of watching Hong Kong films released between the 1970s and the late 1990s, 21.8% of respondents reported having watched three films or fewer, 27.2% had watched four to six films, 16.4% had watched seven to nine films, and 34.6% had watched 10 films or more. More than 81% of the respondents reported recognizing all 11 Hong Kong movie stars from the list of celebrities in those days. Concerning fandom relating to Hong Kong films in those days, 59.8% of the respondents agreed. About 26.5% reported that they still watched Hong Kong films. Approximately 59% of the respondents expressed an interest in visiting a film set in a domestic region or an overseas country. More details are explained in Table 1.

Factor analyses and reliability tests

In order to identify the underlying dimensionality of the instruments used in this study, including film nostalgia, perceived cultural proximity, and familiarity with Hong Kong due to exposure to films from the "good old days", exploratory factor analyses were conducted using principal component factoring and varimax rotation. First, factor analysis of a nostalgia scale containing 26 items determined that a 5-factor solution in which each factor had an eigenvalue of over 1.0 was appropriate. An examination of a scree plot supported the factor solution. The five extracted domains explained 17.84%, 12.26%, 11.10%, 9.58%, and 6.90% respectively of the variance.

The value for the Kaiser–Meyer–Olkin (KMO) measure of sampling adequacy was .92, and Bartlett's test of sphericity produced a value of 6,578 ($p < 0.001$). These values indicated validation of the factor model and the existence of one or more factors in the factor solution. The factor loadings, which measure the correlation between the observed measurements and the factors, were located between .43 and .85 close to and far exceeding the 0.45 threshold proposed by Comrey and Lee (1992). In terms of internal consistency for the items within each domain, the reliability alpha values for the five factors ranged from .79 to .84. The reliability alpha values for all five factors exceeded the threshold (0.70) recommended by Nunnally (1978). Table 2 reports the results explicitly.

The results of the factor analyses using perceived cultural proximity (six items) and familiarity with Hong Kong (four items) generated the single-factor models. The single-factor solutions for the two constructs accounted for 55.73% and 81.9% of the variance respectively. The values for the KMO measure of sampling adequacy were 0.85 and 0.84 respectively, and Bartlett's test of sphericity

Table 1. Profile of respondents ($N = 610$).

Category	Percentage (%)	Category	Percentage (%)
Gender		*Marital status*	
Male	51.3	Single	13.8
Female	48.7	Married	84.4
		Other	1.8
Age (years)	11.4	*Occupation*	45.4
40–44	23.5	Company employee	22.3
45–49	30.6	Housewife	11.6
50–54	26.4	Self-employed	7.7
55–59		Professional	4.3
		Civil servants	16.4
		Other	
Residence area			
Metropolitan area or large city	62.8		
Medium or small city	32.3		
Countryside or others	4.9		
Category	Percentage (%)	Category	Percentage (%)
Experience of watching 30 famous Hong Kong		*Know a Hong Kong movie star*	
		Jacky Chan	
films	81.5	Bruce Lee	99.5
A Better Tomorrow (1986)	74.3	Chow Yun Fat	99.0
Swordsman II (1992)	63.4	Leslie Cheung	98.5
Police Story (1985)	61.5	Sammo Hung	97.2
Bao Zheng (1994)	58.5	Andy Lau	93.1
The Shaolin Temple (1982)	53.9	Wang Zuxian	97.5
Farewell My Concubine (1993)	53.9	Chow Sing Chi	91.6
Shaolin Soccer (2001)	53.0	Leung Chiu Wai	84.9
A Chinese Ghost Story (1987)	51.8	Maggie Chung	84.1
Comrades: Almost a Love Story (1996)	50.3	Aron Kwok	81.6
		Level of current fandom for Hong Kong films	
		Strongly disagree	
Armor of God (1986)	21.8	Disagree	66.2
Number of Hong Kong films watched released	27.2	Neutral	4.9
between the 1970s and the late 1990s	16.4	Agree	30.2
3 films or fewer	34.6	Strongly agree	38.4
4–6 films			24.4
7–9 films			2.1
10 films or more			
Level of fandom for Hong Kong films from the 1970s to the late 1990s			
Strongly disagree	1.3	*Experienced a visit to a film set in a domestic*	40.7
Disagree	8.0	*region or overseas country*	59.3
Neutral	30.8	No	
Agree	48.7	Yes	
Strongly agree	11.1		

produced values of 1,306 ($p < 0.001$) and 1,915 ($p < 0.001$) respectively. The results validated the factor models and the existence of one factor in the factor solutions. All of the factor loadings on both constructs were greater than .90. The reliability alpha values of the factors were .84 and .93 respectively, indicating good internal consistency between the items within each construct. These results are shown in Table 3.

Regression analysis to explain familiarity with the film destination, intention to engage in film nostalgia tourism, and influence of nostalgia films on decision-making on the film destination

Before running multiple regression analyses using five film nostalgia domains as independent variables, multicollinearity, which can occur in cases in which independent variables are highly correlated, was checked. This study adopted variance inflation factor (VIF) values as criteria to identify the existence of multicollinearity. Because the values lower than 2.73 in all regression models were lower than the threshold point of 10 (Field, 2000), the equation models were not at risk of multicollinearity.

The results of the regression analysis to explain familiarity with the film destination due to exposure to "good old days" Hong Kong films, revealed an adjusted R^2 of .49, which indicates that the five film nostalgia domains explained 49% of the variance in predicting familiarity with the film destination after watching "good old days" Hong Kong films. Specially, Nostalgia 2 (memory of film content and movie stars; $\beta = .084$, $p < .05$), Nostalgia 3 (reminiscence of mimicking; $\beta = .246$, $p < .000$), Nostalgia 4 (memory of film backdrops; $\beta = .324$, $p < .000$), and Nostalgia 5 (memory of Hong Kong history and culture; $\beta = .181$, $p < .000$) were strong contributors to the explanation of familiarity with the film destination.

The regression model to explain the intention to engage in a film nostalgia tourism tour due to exposure to Hong

Table 2. Factor analysis of film nostalgia about Hong Kong films.

Domains and items	Factor loading	Communality	Mean
Domain 1 (Yearning for advanced society)			
(α = 0.84, eigenvalue = 4.64, explained variance = 17.84%)			
I believed Hong Kong's political and social system looked stable.	.74	.63	2.85
I believed the industrial products featured in the films were technologically advanced.	.74	.62	3.23
I believed Hong Kong's economic development was advanced.	.72	.61	3.43
I envied Hong Kong society as shown in the films.	.68	.53	3.00
I believed that Hong Kong society, as shown in the films, was vivid and dynamic.	.65	.55	3.60
I felt that the society depicted in the Hong Kong films was free and not restricted.	.55	.41	3.38
The Hong Kong food featured in the films looked tasty.	.54	.46	3.43
Grand mean = 3.28			
Domain 2 (Memory of films' content and movie stars)			
(α = 0.79, eigenvalue = 3.19, explained variance = 12.26%)			
The fighting/martial arts scenes in the Hong Kong action movies were interesting.	.78	.66	4.03
The storylines in the Hong Kong films were intriguing.	.70	.56	3.94
The fight scenes featured in the films were memorable.	.67	.58	3.94
The special effects in the films (e.g. flying in a Kung Fu movie) were wonderful.	.65	.52	3.68
The fashion style of the Hong Kong film stars was stylish and beautiful.	.53	.49	3.53
The Hong Kong film stars were attractive.	.47	.42	3.85
Grand mean = 3.83			
Domain 3 (Reminiscence of mimicking)			
(α = 0.82, eigenvalue = 2.89, explained variance = 11.10%)			
I repeated or followed the lines that the actors/actresses spoke.	.79	.71	3.12
I mimicked the actions of the Hong Kong film actors/actresses such as performing martial arts, wearing sunglasses, smoking cigarettes, and putting on make-up.	.76	.66	3.19
I wanted to learn the martial arts featured in the films.	.71	.67	3.46
I wanted to have the brand products featured in the films.	.61	.64	2.91
Hong Kong films were the topic of talk or gossip among friends.	.45	.65	3.77
Grand mean = 3.29			
Domain 4 (Memory of film backdrops)			
(α = 0.79, eigenvalue = 2.49, explained variance = 9.58%)			
I liked the background music in the films.	.58	.51	3.85
The alleys, traditional markets, restaurants, and buildings featured in the films were memorable.	.54	.60	3.22
The scenes of streets at night featured in the films were shiny and vivid.	.51	.57	3.59
The background scenery featured in the Hong Kong films was beautiful.	.44	.48	3.60
The scenes in the Hong Kong films showed us that Hong Kong was an international city.	.43	.42	3.73
The Hong Kong harbor as featured in the films was fantastic.	.43	.53	3.50
Grand mean = 3.58			
Domain 5 (Memory of Hong Kong history and culture)			
(α = 0.74, eigenvalue = 1.79, explained variance = 6.90%)			
I could understand Hong Kong history through the historical stories shown in the films.	.85	.78	3.08
I could understand the Hong Kong/Chinese customs and culture featured in the Hong Kong films.	.81	.75	3.42
Grand mean = 3.25			

Items were measured on a 5-point Likert scale (*strongly disagree* (1), *neutral* (3), *strongly agree* (5)).

Table 3. Factor analysis of perceived cultural proximity and familiarity with a film nostalgia destination.

Domain and items	Factor loading	Communality	Mean
Perceived cultural proximity (α = 0.84, eigenvalue = 3.34, explained variance = 55.73%)	.81	.60	3.09
I think that the cultural traditions between the two countries are similar.			
I think the stories of Hong Kong films are understandable to me as a Korean person.	.81	.48	3.50
I think that attitudes toward individuals and society in the two nations are similar.	.78	.65	3.27
I think that thoughts about how to achieve harmony within an organization or a group are similar in the two countries.	.72	.65	3.27
I think that the food is similar in the two countries.	.69	.44	2.85
I think that methods for recognizing and solving interpersonal conflict are similar in the two countries.	.66	.52	3.24
Grand mean = 3.20			
Domain and items	Factor loading	Communality	Mean
Familiarity with a film destination (α = 0.93, eigenvalue = 3.28, explained variance = 81.9%)			
I feel closer to Hong Kong because of past Hong Kong films.	.92	.82	3.44
I feel more familiar with Hong Kong because of past Hong Kong films.	.91	.84	3.51
I feel more favorable about the image of Hong Kong because of past Hong Kong films.	.90	81	3.54
I feel more accustomed to Hong Kong because of past Hong Kong films.	.90	.81	3.52
Grand mean = 3.50			

Items were measured on a 5-point Likert scale (*strongly disagree* (1), *neutral* (3), *strongly agree* (5)).

Kong films from the "good old days" revealed an adjusted R^2 value of .27, which means that the five film nostalgia domains account for 27% of the variance in the intention to engage in film nostalgia tourism to Hong Kong. In particular, Nostalgia 3 (reminiscence of mimicking; $\beta = 0.138$, $p < 0.000$), Nostalgia 4 (memory of film backdrops; $\beta = 0.315$, $p < 0.000$), and Nostalgia 5 (memory of Hong Kong history and culture; $\beta = 0.113$, $p < 0.000$) were strong predictors with regard to explaining the intention to participate in film nostalgia tourism to Hong Kong.

The regression analysis to explain the influence of "good old days" Hong Kong films on deciding to travel to Hong Kong generated an adjusted R^2 value of .17, which can be interpreted to indicate that the five film nostalgia domains explain 17% of the variance in the influence of "good old days" Hong Kong films on the decision to travel to Hong Kong. Among the five independent variables, Nostalgia 3 (reminiscence of mimicking; $\beta = 0.147$, $p < 0.000$), Nostalgia 4 (memory of film backdrops; $\beta = 0.232$, $p < 0.000$), and Nostalgia 5 (memory of Hong Kong history and culture; $\beta = 0.098$, $p < 0.000$) were strong contributors regarding an explanation of the influence of "good old days" Hong Kong films on the decision to travel to Hong Kong. As a result, film nostalgia is an influential predictor of film nostalgia tourism intention and decision-making regarding travel to film tourism destinations. The results of the regression analyses are illustrated in Table 4.

Cluster analysis

According to the results of the regression analyses, film nostalgia was efficacious in explaining the intention to engage in film nostalgia tourism. Cluster analysis was adopted to explore the distinctiveness of the film nostalgia market segments. Figure 2 presents the results of the K-means clustering analysis, which used the mean values of the five factors as the results of explanatory factor analysis for film nostalgia. The four-cluster solution was most interpretable and acceptable in the even distribution of membership in each cluster. Clusters 1, 2, 3, and 4 included 26.7%, 19.8%, 23.8%, and 29.7% respectively.

Cluster 1 showed a medium level in nostalgia domains 1–4 but a low level in nostalgia 5 (memory of Hong Kong history and culture). Cluster 1 was thus labeled "group with a medium level of film nostalgia". Cluster 2 was featured as a cohort of the highest level on all five film nostalgia domains, whereas Cluster 3 had the lowest level in all five film nostalgia domains. Thus Cluster 2 was named "group with a high level of film nostalgia", whereas Cluster 3 was labeled "group with a low level of film nostalgia". Cluster 4 showed a low level in nostalgia 3 (reminiscence of mimicking), a high level in nostalgia 5 (memory of Hong Kong history and culture), and a medium level in the other three domains. As a result, Cluster 4 was named "group with a mixed level of film nostalgia".

Difference in involvement, perceived image, and familiarity and perceived cultural proximity among four clusters

A series of one-way analysis of variance (ANOVA) was undertaken to assess the mean differences in involvement, image, familiarity, and perceived cultural proximity that film audiences perceived among the four film nostalgia clusters. Detailed outcomes are reported in Table 5. Significance was established for all seven psychological

Table 4. Results of regression analysis explaining familiarity, intention, and influence.

Dependent variable (Familiarity with the film destination): *Adj-R^2* = 0.49 (F = 117.44, p = 0.000)	Beta (β)	*T*-value	*P*-value
Nostalgia 1 (Yearning for advanced society)	.062	1.54	.124
Nostalgia 2 (Memory of film content and movie stars)	.084	2.11*	.035
Nostalgia 3 (Reminiscence of mimicking	.246	6.64***	.000
Nostalgia 4 (Memory of film backdrops)	.324	6.76***	.000
Nostalgia 5 (Memory of Hong Kong history and culture)	.181	5.55***	.000
Dependent variable (Intention to participate in film nostalgia tour to the film destination): *Adj-R^2* = 0.27 (F = 46.68, p = 0.000)	Beta (β)	*T*-value	*P*-value
Nostalgia 1 ((Yearning for advanced society)	.087	1.79	.074
Nostalgia 2 (Memory of film content and movie stars)	.003	.062	.951
Nostalgia 3 (Reminiscence of mimicking	.138	3.13**	.002
Nostalgia 4 (Memory of film backdrops)	.315	5.52***	.000
Nostalgia 5 (Memory of Hong Kong history and culture)	.113	2.90**	.004
Dependent variable (Influence of "good old days" Hong Kong films on deciding a travel to the film destination): *Adj-R^2* = 0.17 (F = 26.74, p = 0.000)	Beta (β)	*T*-value	*P*-value
Nostalgia 1 (Yearning for advanced society)	.091	1.77	.077
Nostalgia 2 (Memory of film content and movie stars)	.042	.836	.403
Nostalgia 3 (Reminiscence of mimicking	.147	3.12**	.002
Nostalgia 4 (Memory of film backdrops)	.232	3.81***	.000
Nostalgia 5 (Memory of Hong Kong history and culture)	.098	2.36**	.018

$^*p < 0.05$, $^{**}p < 0.01$, $^{***}p < 0.001$. Variance Inflation Factor (VIF) < 2.73.

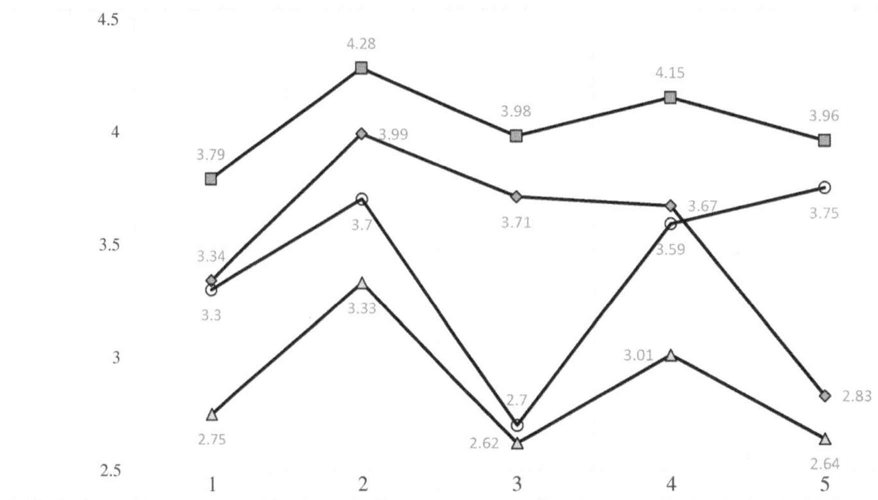

	1	2	3	4	5
Cluster 1	3.34	3.99	3.71	3.67	2.83
Cluster 2	3.79	4.28	3.98	4.15	3.96
Cluster 3	2.75	3.33	2.62	3.01	2.64
Cluster 4	3.3	3.7	2.7	3.59	3.75

Figure 2. Results of cluster analysis using film nostalgia factors.
Cluster 1 (group with a medium level of film nostalgia), Cluster 2 (group with a high level of film nostalgia), Cluster 3 (group with a low level of film nostalgia), Cluster 4 (group with mixed levels of film nostalgia).

Table 5. One-way analysis of variance for comparison of involvement, image, familiarity, and perceived cultural proximity for different film nostalgia clusters.

	Clusters					
Psychological and behavioral involvement	1	2	3	4	*F*-value	*P*-value
Psychological involvement						
Hong Kong films were interesting.	3.84b	4.07c	3.16a	3.77b	75.95	.000
Watching Hong Kong films was an important part of my life.	3.24b	3.58c	2.53a	3.14b	56.84	.000
Watching Hong Kong films tells others who I am.	2.98b	3.33c	2.37a	2.86b	48.46	.000
Hong Kong films brought me joy.	3.74c	3.95d	2.92a	3.58b	68.64	.000
Hong Kong films offered import meaning in my life.	2.99b	3.29c	2.28a	2.83b	44.29	.000
Watching Hong Kong films was part of revealing myself (who I am) to others.	2.96b	3.35c	2.23a	2.80b	46.78	.000
Hong Kong films made me develop curiosity about other countries or the world.	3.51b	3.87c	2.74a	3.50b	65.51	.000
Behavioral involvement						
I like to watch Hong Kong films.	3.71b	4.00c	3.03a	3.56b	53.58	.000
I like to purchase Hong Kong film CDs or music discs.	2.90b	3.17c	2.15a	2.72b	44.19	.000
I like to talk about Hong Kong films.	3.49c	3.77d	2.64a	3.25b	60.87	.000
I am interested in updates of information on Hong Kong films.	3.12b	3.51c	2.43a	3.02b	43.51	.000
I look up whether a new Hong Kong film has been released.	2.87b	3.30c	2.21a	2.71b	40.21	.000
I search for any news related to Hong Kong film stars.	3.15b	3.51c	2.33a	3.02b	59.50	.000
Perceived image						
I felt that Hong Kong was an international city.	4.01c	4.26d	3.41a	3.86b	45.19	.000
I believed that Hong Kong was economically developed.	3.81b	4.16c	3.32a	3.72b	41.77	.000
I believed that Hong Kong was an international tourist city.	3.92b	4.28c	3.48a	3.87b	39.70	.000
I believed that Hong Kong was a city where the film industry was well developed.	3.93b	4.26c	3.46a	3.82b	39.86	.000
I believed that products sold in Hong Kong were trustworthy.	3.20b	3.45c	2.70a	3.09b	30.37	.000
I believed that Hong Kong citizens adhere to social morality.	2.91b	3.26c	2.50a	3.04b	30.81	.000
I believed that Hong Kong was full of romance.	3.55b	3.88c	2.94a	3.62b	58.44	.000
I believed that Hong Kong was full of vigor and energy.	3.82b	4.11c	3.17a	3.77b	76.01	.000
I believed that Hong Kong citizens were noisy in public places.	3.63b	3.85c	3.40a	3.56b	9.48	.000
I believed that Hong Kong was crowded with people and cars.	3.63b	3.86c	3.39a	3.62b	11.06	.000
I believed that Hong Kong was violent due to the crimes featured in the films.	3.52b	3.83c	3.28a	3.62b	13.20	.000
I believed that Hong Kong streets were dirty and disorderly.	3.16a	3.45b	3.08a	3.19a	6.94	.000
I believed that Hong Kong restaurants were unsanitary and untidy.	3.06a	3.28b	2.97a	3.01a	5.44	.000
Familiarity	3.57b	4.07c	2.82a	3.61b	120.02	.000
Perceived cultural proximity	3.24b	3.54c	2.80a	3.27b	53.50	.000

a, b, c, and d indicate the source of significant differences at the 0.001 level (a < b < c < d).

involvement items, all six behavioral involvement items, all 13 perceived image items, one familiarity domain, and one perceived cultural proximity domain. The results are almost consistent in that Cluster 2 showed the highest mean score, followed by Clusters 1, 4, and 3.

Regarding perceived image, Cluster 2 ("group with a high level of film nostalgia") had the highest value for a favorable image of film backdrops, such as the perception of an economically developed global dynamic or romantic city. However, it is interesting that they also clearly had unfavorable images of unsanitary restaurants and noisy, violent, and unclean streets. However, compared with other clusters, Cluster 3 ("group with a low level of film nostalgia") showed the most favorable image of negative image items and indicated the most unfavorable image of the positive image items.

Difference in preferred activities at the film nostalgia destination among four clusters

Cluster 2 ("group with a high level of film nostalgia") showed the highest mean values on all items. However, Cluster 3 ("group with a low level of film nostalgia") showed the lowest level of preference for tourism-related film nostalgia. Cluster 1 ("group with a medium level of film nostalgia") revealed the second-highest level of interest in film nostalgia tourism products or programs. Cluster 4 ("group with a mixed level of film nostalgia") showed a pattern similar to those of Cluster 1 in terms of the magnitude of mean scores. However, the mean scores of the members of Cluster 4 were significantly higher than those of the members of Cluster 3. The results are presented in Table 6.

Difference in preferred tour programs or products to induce film nostalgia tourists among four clusters

In the comparison of preferred tour programs or products to attract film nostalgia tourism among four clusters, the results of one-way ANOVA tests showed significance at the .001 level on all 11 items. The patterns in the location of sources of significant differences and the magnitude of the mean scores among clusters were similar. A cohort with a high level of film nostalgia (Cluster 2) showed the highest mean scores, whereas those with the least involvement with film nostalgia (Cluster 3) reported the lowest mean scores. The second-highest mean scores were noticed on Clusters 1 and 4. The outcomes are shown in Table 7.

Difference in sociodemographic and film fandom-related characteristics among four clusters

A series of chi-square tests were conducted to assess statistically significant levels of association between the four clusters and the selected sociodemographic and film fandom-related indicators. Significant differences were observed at the .05, .01, and .001 levels on age, sex, occupation, previous and current film fandom, and experience with visiting film sets. The results are reported in Table 8.

With regard to age, the proportions of participants 40–44 years of age in the "group with a high level of film nostalgia" (Cluster 2) and the "group with a low level of film nostalgia" (Cluster 3) were 35.5% and 39.3% respectively. The "group with little memory of Hong Kong history and culture" (Cluster 1) had the highest percentage of members 40–44 years of age (42.3%). The

Table 6. One-way analysis of variance tests for comparison of preferred activities at the film nostalgia destination for different film nostalgia clusters.

	Clusters					
Preferred activities	1	2	3	4	*F*-value	*P*-value
I'd like to visit alleys, traditional markets, buildings, and restaurants that were shown in the films in those "good old days".	3.52b	4.04c	2.88a	3.46b	58.24	.000
I'd like to visit indoor/outdoor film sets.	3.58c	4.01d	2.79a	3.37b	57.41	.000
I'd like to buy things that were shown in the films in those "good old days".	3.15c	3.47d	2.43a	2.92b	45.99	.000
I'd like to visit the film studios that developed the films.	3.29c	3.79d	2.57a	3.10b	56.20	.000
I'd like to take part in "fan-sign" meetings that film stars from those "good old days" attend.	3.37c	3.78d	2.52a	3.05b	63.48	.000
I'd like to meet the movie stars in those "good old days".	3.75c	4.18d	2.97a	3.45b	64.84	.000
I'd like to visit the Hong Kong Film History Museum.	3.45b	4.03c	2.66a	3.30b	74.12	.000
I'd like to taste the Hong Kong (Cantonese) food featured in the films in those "good old days".	3.60b	4.17c	2.89a	3.46b	60.83	.000
I'd like to travel to Hong Kong to trace my memories of the films in those "good old days".	3.69b	4.18c	3.03a	3.67b	52.60	.000
I'd like to talk with other tourists who watched Hong Kong films in those "good old days" to share reminiscences.	3.29c	3.76d	2.56a	3.07b	56.00	.000
I'd like to go to the restaurant and eat food where Hong Kong actors/actresses visited in the films.	3.55b	4.03c	2.79a	3.45b	62.10	.000

a, b, c, and d indicate the source of significant differences at the 0.001 level (a < b < c < d).

Table 7. One-way analysis of variance tests for comparison of preferred tour programs or products to induce film nostalgia tourists for different film nostalgia clusters.

Preferred tour programs or products	Clusters 1	2	3	4	F-value	P-value
During a visit to Hong Kong, a tour program that interprets the famous Hong Kong films in those "good old days" should be developed.	3.53b	3.95c	2.84a	3.41b	58.28	.000
Fashions or accessories worn by Hong Kong film stars should be developed as products for tourists.	3.19c	3.55d	2.59a	3.03b	47.69	.000
Mobile products of Hong Kong film stars or film content, pictures, and text messages in Korean that can be downloaded should be developed.	3.13b	3.49c	2.50a	3.01b	43.25	.000
Programs that explain Hong Kong films should be developed by Korean television networks.	3.06b	3.60c	2.55a	3.13b	46.23	.000
"Fan-sign" events that Hong Kong film stars attend should be organized.	3.19c	3.58d	2.54a	2.99b	49.60	.000
Internet portal sites where Hong Kong films are introduced should be developed.	3.18b	3.63c	2.60a	3.18b	44.95	.000
Travel packages that include visiting places featured in films should be developed.	3.47b	4.00c	2.77a	3.38b	65.39	.000
Themed music concerts featuring Hong Kong film original sound track (OST) or background music should be held.	3.39b	3.73c	2.70a	3.28b	47.92	.000
During a visit to Hong Kong, a travel package that includes visits to Hong Kong film sets should be developed.	3.45b	3.88c	2.73a	3.32b	59.92	.000
During a visit to Hong Kong, a travel package that includes visits to the Hong Kong Film History Museum should be developed.	3.41b	3.84c	2.68a	3.29b	64.34	.000
Travel guidebooks that introduce Hong Kong film scenes and locations should be published.	3.40b	3.85c	2.76a	3.29b	51.37	.000

a, b, c, and d indicate the source of significant differences at the 0.001 level (a < b < c < d).

"group with mixed levels of film nostalgia" (Cluster 4) had a high percentage of members 51 years of age or older (56.3%).

Concerning gender, the "group with a high level of film nostalgia" (Cluster 2) consisted mostly of males (74.8%), as did the "group with a low level of film nostalgia" (Cluster 3; 66.9%). Cluster 1 had more men (54.5%) than women, whereas Cluster 4 had more women (55.2%) than men. Regarding occupation, the group with a high level of film nostalgia (Cluster 2) had the highest percentage of company employees (55.8%), and the "group with a low level of film nostalgia" had the lowest percentage of company employees (36.5%). The highest percentage (31.5%) of housewives was in the "group with a low level of film nostalgia".

In response to a question about fandom regarding "good old days" Hong Kong films, the "group with a high level of film nostalgia" (Cluster 2) showed the highest percentage of agreement on the question. Regarding the number of Hong Kong films watched from the 1970s to late 1990s, Cluster 2 had the highest percentage of members who answered "10 films or more" (48.5%), and Cluster 3 had the lowest percentage in that category (21.44%). Regarding the level of influence of "good old days" Hong Kong films on the decision to travel to Hong Kong, Cluster 2 showed the highest percentages in the "agree" and "strongly agree" categories (33.1%). Cluster 3 had the lowest percentage in these categories (3.5%).

The "group with a low level of film nostalgia" (Cluster 3) had the lowest agreement on current audience of Hong Kong films. Similarly, the "group with a high level of film nostalgia" (Cluster 2) revealed the highest percentage of agreement with experience visiting domestic or overseas film sets, and the lowest percentage was seen in the "group with a low level of film nostalgia (Cluster 3).

Difference in film nostalgia tourism characteristics among four clusters

The outcomes of the correspondence analyses provided graphic information on the associations of the film nostalgia clusters (row variable) and preferred film nostalgia tourism characteristics (column variables) in a two-dimensional plot. The proximity between a pair of points for column and row variables was used to interpret the strength of their underlying relationship; that is, the closer the two points, the stronger the relationship between them.

In the correspondence analysis between the five clusters and preferred duration of film nostalgia tours, two principal components collectively accounted for 96.9% of the variance, with 89.8% (singular value = .15) accounted for by the first dimension and 7.1% (singular value = .04) accounted for by the second dimension (Figure 3). According to the results, the members of Clusters 1 and 4 preferred to stay for one night, and those in Cluster 2 preferred to stay for two nights.

With regard to the explained proportion of inertia in the correspondence analysis between the four clusters and the preferred film-related gift item, the first two principal components accounted for 98.2% of the

Table 8. Chi-square tests for comparison of the sociodemographic and film fandom-related characteristics for different film nostalgia clusters.

Sociodemographic and film fandom-related characteristics	Clusters 1	2	3	4	Chi-square	P-value
Sociodemographic characteristics						
Age (years)						
40–44	42.3	35.5	39.3	24.9	17.61	.040
45–50	19.0	23.1	19.3	18.8		
51–54	20.9	19.8	23.4	29.8		
55 or older	17.8	21.5	17.9	26.5		
Gender						
Male	54.5	74.8	33.1	44.8	62.99	.000
Female	45.5	13.8	66.9	55.2		
Marital status						
Single	14.6	12.4	18.2	11.3	3.44	.328
Married	85.4	87.6	81.8	88.7		
Occupation						
Company employee	51.2	55.8	36.5	40.0	46.68	.000
Independent businessman	6.6	16.0	11.6	11.0		
Civil servant	5.8	5.5	1.5	4.1		
Housewife	23.1	9.2	31.5	24.8		
Professional	5.8	6.7	11.0	6.2		
Other (no job, agricultural and fishery, transportation)	7.4	6.7	7.2	13.8		
Residence area						
Large city	59.5	67.8	60.7	64.1	3.72	.715
Medium city	35.0	29.8	33.1	30.9		
Rural area	5.5	2.5	6.2	5.0		
Film fandom-related characteristics						
Fandom of "good old days" Hong Kong films						
Strongly disagree	1.1	1.2	2.1	1.6	139.89	.000
Disagree	1.1	1.2	19.3	8.8		
Neutral	26.4	11.6	44.8	36.5		
Agree	57.1	58.7	30.3	49.2		
Strongly agree	14.7	28.1	3.5	3.9		
Number of Hong Kong films watched between the 1970s and the late 1990s						
3 films or fewer	15.7	12.9	37.2	21.5	63.03	.000
4–6 films	18.2	23.3	29.0	35.4		
7–9 films	19.0	15.3	12.4	18.8		
10 films or more	47.1	48.5	21.4	24.3		
Do you watch Hong Kong films nowadays?						
Strongly disagree	1.1	4.1	11.0	3.9	80.05	.000
Disagree	25.2	25.6	40.0	29.8		
Neutral	43.6	29.8	39.3	38.7		
Agree	29.4	31.4	9.0	27.6		
Strongly agree	1.1	9.1	.7	.0		
If you decide travel to Hong Kong, is this influenced by "good old days" Hong Kong films?						
Strongly disagree	7.4	6.6	33.1	8.8	106.96	.000
Disagree	40.5	25.6	40.0	43.1		
Neutral	38.7	34.7	23.4	35.4		
Agree	12.3	29.8	2.1	12.2		
Strongly agree	1.2	3.3	1.4	0.6		
Did you visit domestic or overseas film sets?						
No	34.7	34.4	53.1	40.3	13.77	.003
Yes	65.3	65.6	46.9	59.7		

All numbers represent percentages.

variance, with 80.3% (singular value = .15) accounted for by the first dimension and 17.9% (singular value = .07) accounted for by the second dimension (Figure 4). As Figure 4 shows, the respondents in Cluster 1 showed a preference for film booklets and handkerchiefs, whereas those in Cluster 2 preferred T-shirts from the film locations. The members in Cluster 3 preferred postcards from the film locations, whereas those in Cluster 4 preferred digital versatile discs (DVDs), compact discs (CDs), and handkerchiefs.

The first principal component in the correspondence analysis between the four clusters and the preferred food in a film nostalgia tourism destination explained 97.0% of the variance, with 77.5% (singular value = .13) accounted for by the first dimension and 19.5% (singular value = .07) accounted for by the second dimension (Figure 5). According to the outcomes of the analysis, the members of Clusters 1 and 2 indicated a preference for local Hong Kong food, whereas those in Cluster 4 preferred Sichuan food and those in Cluster 3 preferred Beijing food.

In the correspondence analysis of factors identifying the relationship between the four clusters and the preferred types of film nostalgia travel, two principal components explained 100% of the variance, with 78.2% (singular value = .11) accounted for by the first dimension and 21.8% (singular value = .06) accounted for by the second dimension (Figure 6). The respondents in Cluster 2 preferred individual tours, whereas those in Cluster 3 preferred package tours. The members of Clusters 1 and 4 preferred a combination of individual and package tours.

Discussion and conclusions

Contributions and implications

The findings of this study contribute to the current film tourism literature theoretically and practically. The four key theoretical contributions and implications resulting from this study are discussed in full in the following sections.

First, the possible potential role nostalgia plays in understanding film tourism. This phenomenon has been discussed in various contexts; for example, in relation to imagination, emotional involvement, and place attachment (Bandyopadhyay, 2008; Kim, 2012a; Reijinders, 2016; Rittichainuwat & Rattanaphinanchai, 2015; Roesch, 2009). There have been few previous studies, if any, that have treated nostalgia as an underpinning motivation for the creation of film tourism. The results of this study indicate that an understanding of film tourism should start with a thorough examination of the concept of nostalgia, with particular regard to the multidimensional nature of this concept, and the variety of relationships it has with other relevant concepts already established in the film tourism literature (e.g. involvement, destination image, familiarity, and

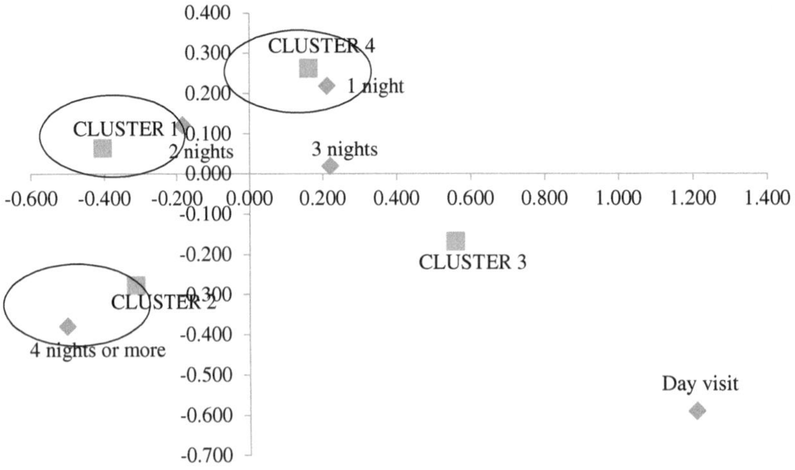

Figure 3. Correspondence analysis of film nostalgia clusters and duration of stay.

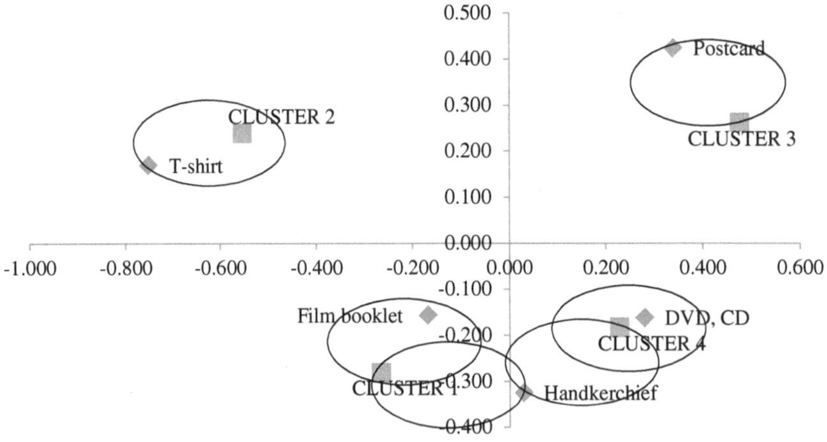

Figure 4. Correspondence analysis of film nostalgia clusters and preferred film-related gifts.

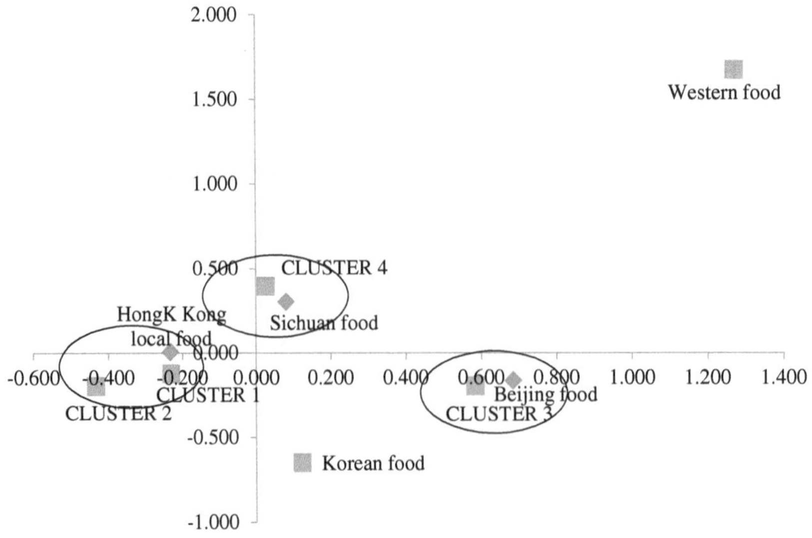

Figure 5. Correspondence analysis of film nostalgia clusters and preferred food during film tourism.

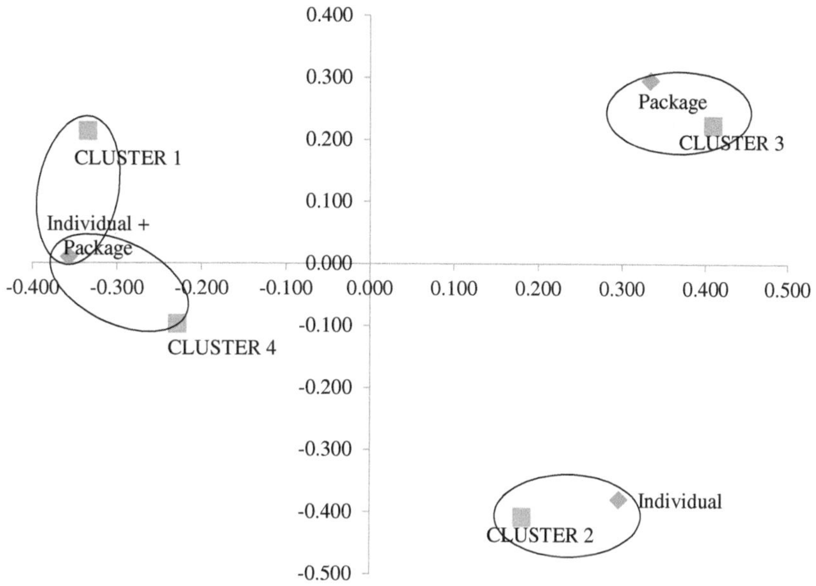

Figure 6. Correspondence analysis of film nostalgia clusters and preferred film tourism type.

perceived cultural proximity). Such complexity indicates that further studies, to empirically examine these underlying relationships that exist between the above variables, would be pertinent.

The findings of this study also have a particular contribution to make toward furthering understanding of film tourism. Given that the relationship between nostalgia and familiarity has only been theoretically discussed in the relevant literature, empirical testing of the dimensions of nostalgia and familiarity in the context of film tourism could establish any causal relationships. There were three film nostalgia domains (reminiscence of mimicking, memory of film backdrops, and memory of Hong Kong history and culture) that proved significant in explaining three key elements of film tourists' behavior. These key elements were familiarity with Hong Kong after watching Hong Kong films, the intention to participate in a film nostalgia tour to film destinations, and the decision to travel to film destinations. Some film audiences who harbored more nostalgic sentiments about imitating the actions of film stars, visiting film backdrops, and learning about Hong Kong history and culture (the so-called "specific or purposeful" potential film tourists) revealed a stronger familiarity with Hong Kong as well as a stronger intention to choose Hong Kong as a potential film tourism destination. It is important to note that memory of film content and movie stars was not significant in explaining the intention to engage in film tourism to Hong Kong and the decision to travel to film locations. This finding can probably be explained by the fact that most Hong Kong films screened during that period were dominated by elements of film noir (portraying gangsters' backstreet lives, crime, drugs, violence, and murder without reason (H. Lee, 2006)), and although Korean audiences enjoyed watching them at the time, the content and actors of the films did not lead to the development of strong memories in the audiences. This finding is inconsistent with previous studies, which highlighted the importance of stories and characters or actors in provoking stronger emotional attachment and thus prolonged memory (Kim, 2012a; Kim & Assaker, 2014; Reijnders, 2016).

Second, it has been documented that different sets of specific motivations, interests, expectations, and experiences are demonstrated by different groups of film tourists (e.g. specific/purposeful, incidental, casual, or serendipitous film tourists) (Croy & Heitmann, 2011; Kim, 2012a; Macionis & Sparks, 2009). However, it should be noted that no previous studies have conducted a cluster analysis to empirically classify different groups of prospective film tourists and examine their underlying characteristics. The findings of this study fill the above gap by confirming that the four identified film nostalgia tourism clusters or groups have distinctive characteristics. A follow-up study on the results from the current study is expected to have the benefit of defining a segmentation variable for film tourists more clearly. A future study should also apply this concept to other film tourism destinations to verify whether film nostalgia could also explain characteristics of film tourists. This would be particularly useful for those in the hospitality and tourism industries who may consider developing film tourism-themed attractions, products, merchandise, and tour programs. This final point will be discussed in more detail later in this section.

Third, the highest level of perceived cultural proximity was found in the "group with a high level of film nostalgia" (Cluster 2). This indicates that audiences with a closer perceived cultural proximity to Hong Kong were also those that had a high level of film nostalgia. These results correspond with those of previous studies, where preference for a film and film tourism to a featured destination was determined by perceived cultural proximity (Bandyopadhyay, 2008; Burch, 2002; Huang, 2011; Su et al., 2011). This important point implies that film tourism may be more prevalent among audiences within similar cultural boundaries to a film's country of origin because the stories, body language, clothing styles, and props convey similar cultural meanings (H. Lee, 2006; Su et al., 2011).

The fourth point is that, among the five nostalgia domains, reminiscence of mimicking was the most influential in explaining familiarity with film destinations and the intentions to visit these places. This finding is understandable as behavioral involvement prolongs the memory more than psychological involvement (Acharya, Paudel, & Hatch, 2009; Holak, 2013; Holbrook, 1993). Those who imitate or re-enact the poses, clothing styles, accessories, or language of actors must be highly involved with the films and are more likely to trace their memories back to the "good old days" by experiencing a film destination. This is similar to the findings of study.

This study also has practical and managerial implications for developing market segmentations that identify characteristics of potential film tourists based on film nostalgia cohorts, given that the efficacy of film nostalgia as a facilitator of film tourism has been clearly demonstrated. Table 9 shows differences in diverse characteristics among the four film nostalgia clusters.

First, the "group with a high level of film nostalgia" (Cluster 2) would be more likely to stay longer, prefer individual tours, enjoy tasting local Hong Kong food, and purchase memorabilia or merchandise. This specific film nostalgia tourism group would be more inclined to want to visit sacred or unknown film places, seek out and re-enact their memories, and experience authentic Hong Kong food as portrayed in their favorite Hong Kong films from the 1970s to the 1990s.

These findings suggest that the travel industry could develop a travel itinerary that includes reminiscence-tracing film locations displaying props such as books, furniture, telephones, analogue products, shoes, hats, and lights from the "good old days". Such itineraries would encourage tourists to re-enact the intriguing film scenes, recite lyrics, or appreciate photos of past film fandom while sharing their memories with people of a similar age. This sharing might in turn stimulate tourists' recollections, which could lead to increased satisfaction and, ultimately, repeat visits or recommendations. In the hospitality sector, local restaurants in particular could benefit from these suggested itineraries by developing menus featuring dishes that older customers remember from the films they saw in the 1970s, 1980s, and 1990s, and are likely to want to taste again. This could be extended into catering for customers who prefer the retro-fashioned restaurant interiors, designs, and memorabilia seen in movies.

It is also important to identify the members of this high-profile group. Those in the "group with a high level of film nostalgia" tended to be male company workers more often in their forties than in their fifties. It is possible to infer from the study that men were more exposed to Hong Kong films that were mainly action movies when they were in their twenties. It makes sense that male audiences liked Hong Kong films that were specific to a particular genre of action or gangster films. These Korean male audiences, whose members are now in their forties, who, according to the results of this study, were (and still are) current fans of Hong Kong films, must be considered as "specific or purposeful film nostalgia tourists". In addition, because most participants in this study were married, they would be accompanied by their spouses and children.

Second, the "group with a low level of film nostalgia" (Cluster 3), which showed the lowest levels of involvement, image familiarity, and perceived cultural proximity, preferred activities of film nostalgia. This group also preferred tour programs or products to draw in film nostalgia tourists. In addition, this group revealed the lowest level of film fandom-related characteristics and preferred a package tour. These group members manifested a preference for buying postcards with photos of film locations and tasting Beijing food, which has little association with the Hong Kong films. This group, that was predominantly female (66.9%), had the lowest percentage of company workers (36.5%) and the highest percentage of housewives (31.5%). They can be characterized as being almost incidental or serendipitous film tourists. They are less influenced by nostalgia films; thus, lacking reminiscences and memories that were an important aspect of "specific or purposeful film tourists" (Kim, 2012a) (identified as Cluster 2 in this study). This finding is understandable because these people were relatively less exposed to Hong Kong films and were not fervent previous or current fans of these films. This group is likely to visit Hong Kong due to other external motivational factors such as shopping.

Third, the "group with a medium level of film nostalgia" (Cluster 1), which had had some previous

Table 9. Summary of comparisons of film nostalgia clusters according to diverse variables.

	Cluster 1 (group with a medium level of film nostalgia) (26.7%)	Cluster 2 (group with a high level of film nostalgia) (19.8%)	Cluster 3 Group with a low level of film nostalgia) (23.8%)	Cluster 4 (group with a mixed level of film nostalgia) (29.7%)
Involvement, image, familiarity, and perceived cultural proximity	• Medium level for involvement • Medium level for both favorable and unfavorable image • Medium level for familiarity • Medium level for perceived cultural proximity	• Highest level for involvement • Highest level for both favorable and unfavorable image • Highest level for familiarity • Highest level for perceived cultural proximity	• Lowest level for involvement • Lowest level for both favorable and unfavorable image • Lowest level for familiarity • Lowest level for perceived cultural proximity	• Medium level for involvement • Medium level for both favorable and unfavorable image • Medium level for familiarity • Medium level for perceived cultural proximity
Preferred activities at the film nostalgia destination	• Medium level for preferred activities	• Highest level for preferred activities	• Lowest level for preferred activities	• Medium level for preferred activities
Preferred tour programs or products to induce film nostalgia tourists	• Medium level for preferred tour programs or products	• Highest level for preferred tour programs or products	• Lowest level for preferred tour programs or products	• Medium level for preferred tour programs or products
Sociodemographic characteristics	• Aged 40–50 years (61.3%) • Male (54.5%) • Company workers (51.2%)	• Aged 40–50 years (58.6%) • Male (78%) • Highest percentage of company workers (55.8%) but lowest percentage of housewives (9.2%)	• Aged 40–50 years (58.6%) • Female (66.9%) • Lowest percentage of company workers (36.5%) but highest percentage of housewives (31.5%)	• Aged 51 years or older (56.3%) • Male (54.5%)
Film fandom-related characteristics	• Previous fandom (Agree and strongly agree (71.8%)) • No. of nostalgia films watched (10 films or more (47.1%)) • Current fandom (Agree and strongly agree (30.5%)) • Influence of nostalgia films on decision making to travel to Hong Kong (Agree and strongly agree (13.5%)) • Experience of visiting film sets (Yes (65.3%))	• Previous fandom (Agree and strongly agree (86.8%)) • No. of nostalgia films watched (10 films or more (48.5%)) • Current fandom (Agree and strongly agree (40.5%)) • Influence of nostalgia films on decision making to travel to Hong Kong (Agree and strongly agree (33.1%)) • Experience of visiting film sets (Yes (65.6%))	• Previous fandom (Agree and strongly agree (33.8%)) • No. of nostalgia films watched (10 films or more (21.4%)) • Current fandom (Agree and strongly agree (9.7%)) • Influence of nostalgia films on decision making to travel to Hong Kong (Agree and strongly agree (3.5%)) • Experience of visiting film sets (Yes (46.9%))	• Previous fandom (Agree and strongly agree (53.1%)) • No. of nostalgia films watched (10 films or more (24.3%)) • Current fandom (Agree and strongly agree (27.6%)) • Influence of nostalgia films on decision making to travel to Hong Kong (Agree and strongly agree (12.8%)) • Experience of visiting film sets (Yes (59.7%))
Preferred film nostalgia tourism characteristics at the film destination	• Prefer to stay for two nights • Prefer to buy film booklets and handkerchiefs • Prefer to taste local food • Prefer a combined individual and package tour type	• Prefer to stay for four nights or more • Prefer to buy T-shirts • Prefer to taste local food • Prefer an individual tour type	• Prefer to buy postcard • Prefer to taste Beijing food • Prefer a package tour type	• Prefer to stay for one night • Prefer to buy DVD and CD. • Prefer to taste Sichuan food • Prefer a combined individual and package tour type

exposure to Hong Kong films and showed a medium level of involvement, image familiarity, and perceived cultural proximity, preferred activities related to film nostalgia. This group also preferred tour programs or products to attract film nostalgia tourists, and featuring film fandom-related characteristics. The group members were predominantly 40–50 years of age (61.3%), male (54.5%), and company workers (51.2%). They indicated a future intention to stay for two nights and on a combination of individual and package tour types, and would enjoy buying film booklets and handkerchiefs as film-related gifts. They also indicated that they would like to taste local Hong Kong food that they had seen in the films.

Although those in Cluster 1 were exposed to Hong Kong films, they had little interest in Hong Kong history and culture. This result may be attributed to the storylines in Hong Kong films that contain elements of film noir action or gangster films and lack historical or cultural content. The members of this cluster preferred to buy inexpensive gifts, such as handkerchiefs with photos of movie stars or film places, film booklets, and food. This group was made up of mainly male company workers in their forties, who have easy and constant exposure to the Internet in their work environment. Given this access to the Internet, video clips that contain mimicking motions can be an effective medium for drawing potential film tourists to specific locations. To capitalize on this use, destination marketers could generate promotional materials in cyberspace.

Fourth, the "group with mixed levels of film nostalgia" (Cluster 4), which showed a medium level of involvement, image familiarity, and perceived cultural proximity, preferred activities of film nostalgia and tour programs or products to attract film nostalgia tourists. In addition, this group preferred film fandom-related characteristics. Most members of Cluster 4 were 51 years of age or older (56.3%) and male (54.5%). This group indicated a preference for staying for one night and for a combination of individual and package tour types. They also liked buying DVDs of nostalgic films, CDs of the songs from the original soundtracks, and local handcraft as gifts.

The level of film nostalgia tourism motivation of the respondents in Cluster 4 was not high, although their motivation level was higher than that of those in Cluster 3. These tourists wanted to stay one night for film tourism programs. Destination marketers need to somehow increase their film tourism motivation level or develop tour programs to cater for their needs. To give an example of how this might work in practice: these tourists prefer to taste Sichuan food in Hong Kong (meaning that they probably have little understanding of what a truly authentic Hong Kong has to offer), so destination marketers could promote the true character and taste of Hong Kong to attract this group. Because of the cohorts' age range, a travel agency needs to promote film nostalgia travel products through newspapers and economy magazines, or by word-of-mouth. In addition, as the population of older people increases, the number of older tourists will increase worldwide. To take advantage of this fact, travel agencies would benefit from targeting older people as a cohort of film tourists because their major motivation for travel is nostalgia (Hsu et al., 2007; Sellicka, 2004).

Finally, it is necessary to analyze the characteristics of these cohorts overall and to select target markets, as four film nostalgia segments have been derived. In this study, the "group with a high level of film nostalgia" (Cluster 2) was revealed to be the most appropriate target market for film nostalgia tourism. Its members showed the highest agreement on all items of preference for activities or products/programs and for psychological and behavioral involvement. They showed a strong desire for film tourism to Hong Kong to help them trace their memories, including film content and stars, mimicking, film backdrops, and Hong Kong history and culture. As discussed above, destination stakeholders should develop products or programs to cater to the specific needs of the target market.

Limitations and suggestions for future research

This study has some limitations. The sample was made up of Korean citizens who had watched Hong Kong films from the 1970s to the late 1990s. As discussed above, there was a strong correlation between perceived cultural proximity and film nostalgia. Future studies could apply the approach used in this study in cross-cultural examinations of other countries in relation to origin of film and film audience (Hudson, Wang, & Gil, 2011; Kim, Agrusa, & Chon, 2014; Kim & O'Connor, 2011).

In addition, this study was limited to subjects who were exposed to Hong Kong films in the 1970s, 1980s, and 1990s. Despite the fact that watching the films in those days was a collective boom phenomenon of social and cultural practice, there would always have been those who did not follow the societal mainstream. A future study could explore the responses of respondents who were not actively exposed to films in those days. As a final point it is worth noting that the potential nostalgia film tourists wanted to stay in Hong Kong for only one or two nights, so there is a need for research into how to make them extend their stay.

Disclosure statement

No potential conflict of interest was reported by the authors.

Funding

The work described in this paper was fully supported by a grant from a Research Grants Council of the Hong Kong Special Administrative Region, China (Project No. PolyU155032/14B).

References

Acharya, R. N., Paudel, K. P., & Hatch, L. U. (2009). Impact of nostalgia and past experience on recreational demand for wilderness. *Applied Economics Letters*, *16*(5), 449–453. doi:10.1080/13504850601032099

Alderman, D. D., Benjamin, S. K., & Schneider, P. P. (2012). Transforming Mount Airy into Mayberry. *Southeastern Geographer*, *52*(2), 212–239. doi:10.1353/sgo.2012.0016

Bandyopadhyay, R. (2008). Nostalgia, identity and tourism: Bollywood in the Indian Diaspora. *Journal of Tourism and Cultural Change*, *6*(2), 79–100. doi:10.1080/14766820802140463

Becken, S., & Gnoth, J. (2004). Tourist consumption systems among overseas visitors: Reporting on American, German, and Australian visitors to New Zealand. *Tourism Management*, *25*(3), 375–385. doi:10.1016/S0261-5177(03)00133-X

Beeton, S. (2010). The advance of film tourism. *Tourism and Hospitality Planning & Development*, *7*(1), 1–6. doi:10.1080/14790530903522572

Buchmann, A., Moore, K., & Fisher, D. (2010). Experiencing film tourism: Authenticity and fellowship. *Annals of Tourism Research*, *37*(1), 229–248. doi:10.1016/j.annals.2009.09.005

Burch, E. (2002). Media literacy, cultural proximity and TV aesthetics: Why Indian soap operas work in Nepal and the Hindu diaspora. *Media, Culture & Society*, *24*(4), 571–579. doi:10.1177/016344370202400408

Carl, D., Kindon, S., & Smith, K. (2007). Tourists' experience of film locations: New Zealand as 'Middle-Earth'. *Tourism Geographies*, *9*(1), 49–63. doi:10.1080/14616680601092881

Chen, C.-C., & Lin, Y.-H. (2012). Segmenting Mainland Chinese tourists to Taiwan by destination familiarity: A factor-cluster approach. *International Journal of Tourism Research*, *14*(4), 339–352. doi:10.1002/jtr.v14.4

Chosun Ilbo. (2016). Reairing of "A better tomorrow" in theatres in celebration of 30-year anniversary. *Chosun Ilbo*. Retrieved January 29, 2016, from http://news.chosun.com/site/data/html_dir/2016/01/29/2016012900051.html

Comrey, A., & Lee, H. (1992). *A first course in factor analysis* (2nd ed.). Hillsdale, NJ: Lawrence Erlbaum.

Connell, J., & Meyer, D. (2009). Balamory revisited: An evaluation of the screen tourism destination-tourist nexus. *Tourism Management*, *30*, 194–207. doi:10.1016/j.tourman.2008.06.001

Couldry, N. (1998). The view from inside the 'simulacrum': Visitors' tales from the set of coronation street. *Leisure Studies*, *17*, 94–107. doi:10.1080/026143698375178

Croy, C. W., & Heitmann, S. (2011). Tourism and film. In P. Robinson, S. Heitmann, & P. U. I. Dieke (Eds.), *Research themes for tourism*. Cambridge, MA: CAB International.

Croy, W. G. (2010). Planning for film tourism: Active destination image management. *Tourism and Hospitality Planning and Development.*, *7*(1), 21–30. doi:10.1080/14790530903522598

Field, A. P. (2000). *Discovering statistics using SPSS for Windows: Advanced techniques for the beginner*. London: Sage.

Frost, W. (2006). Bravehearted Ned Kelly: Historic films, heritage tourism and destination image. *Tourism Management*, *27*, 247–254. doi:10.1016/j.tourman.2004.09.006

Grönlund, K., & Strandberg, K. (2014). Online panels and validity: Representativeness and attrition in the Finnish eOpinion panel. In M. Callegaro, R. Baker, J. Bethlehem, A. S. Göritz, J. A. Krosnick, & P. J. Lavrakas (Eds.), *Online panel research: A data quality perspective* (pp. 86–103). New York, NY: Wiley.

Holak, S. L. (2013). From Brighton beach to blogs: Exploring food-related nostalgia in the Russian diaspora. *Consumption Markets and Culture*, *17*(2), 185–207. doi:10.1080/10253866.2013.776308

Holbrook, M. B. (1993). Nostalgia and consumption preferences: Some emerging patterns of consumer tastes. *Journal of Consumer Research*, *20*(2), 245–256. doi:10.1086/jcr.1993.20.issue-2

Horng, J.-S., Liu, C.-H., Chou, H.-Y., & Tsai, C.-Y. (2012). Understanding the impact of culinary brand equity and destination familiarity on travel intentions. *Tourism Management*, *33*(4), 815–824. doi:10.1016/j.tourman.2011.09.004

Hsu, C. H. C., Cai, L. A., & Wong, K. K. F. (2007). A model of senior tourism motivations: Anecdotes from Beijing and Shanghai. *Tourism Management*, *28*(5), 1262–1273. doi:10.1016/j.tourman.2006.09.015

Hsu, C. H. C., & Song, H. (2013). Destination image in travel magazines: A textual and pictorial analysis of Hong Kong and Macau. *Journal of Vacation Marketing*, *19*(3), 253–268. doi:10.1177/1356766712473469

Huang, S. (2011). Nation-branding and transnational consumption: Japanese-mania and the Korean wave in Taiwan. *Media, Culture and Society*, *33*(1), 3–18. doi:10.1177/0163443710379670

Hudson, S., Wang, Y., & Gil, S. M. (2011). The influence of a film on destination image and the desire to travel: A cross-cultural comparison. *International Journal of Tourism Research*, *13*(2), 177–190.

Hunt, L., & Johns, N. (2013). Image, place and nostalgia in hospitality branding and marketing. *Worldwide Hospitality and Tourism Themes*, *5*(1), 14–26. doi:10.1108/17554211311292411

Kessous, A. (2014). Nostalgia, autobiographical memories and brand strategy: Marketing to the post-World War I generation. *Journal of Brand Strategy*, *3*(2), 148–154.

Kim, S. (2012a). Audience involvement and film tourism experiences: Emotional places, emotional experiences. *Tourism Management*, *33*(2), 387–396.

Kim, S. (2012b). The relationships of on-site film-tourism experiences, satisfaction, and behavioural intentions: From the film-tourism perspective. *Journal of Travel and Tourism Marketing*, *29*(5), 472–484. doi:10.1080/10548408.2012.691399

Kim, S., Agrusa, J., & Chon, K. (2014). The influence of a TV drama on visitors' perception: A cross-cultural study.

Journal of Travel and Tourism Marketing, *31*(3), 536–562. doi:10.1080/10548408.2014.883950
Kim, S., Agrusa, J., Lee, H., & Chon, K. (2007). Effects of Korean television dramas on the flow of Japanese tourists. *Tourism Management*, *28*(5), 1340–1353. doi:10.1016/j.tourman.2007.01.005
Kim, S., & Assaker, G. (2014). An empirical examination of the antecedents of film tourism experience: A structural model approach. *Journal of Travel and Tourism Marketing*, *31*(2), 251–268. doi:10.1080/10548408.2014.873316
Kim, S., & O'Connor, N. (2011). A cross-cultural study of screen-tourists' profiles. *Worldwide Hospitality and Tourism Trends*, *3*(2), 141–158. doi:10.1108/17554211111123005
Kim, S.-J. (2015). Movie diary. *Kyungnam Ilbo*. Retrieved from http://www.knnews.co.kr/news/articleView.php?idxno=1154269
Kim, S. S., & Agrusa, J. (2008). Segmenting Japanese tourists to Hawaii according to tour purpose. *Journal of Travel & Tourism Marketing*, *24*(1), 63–80. doi:10.1300/J073v24n01_05
Kim, S. S., Kim, S., & Heo, C. (2015). Assessment of TV drama/film production towns as a rural tourism growth engine. *Asia Pacific Journal of Tourism Research*, *20*(7), 730–760. doi:10.1080/10941665.2014.927378
Lee, C. (2012). Have magic, will travel: Tourism and Harry Potter's United Kingdom. *Tourist Studies*, *12*(1), 52–69.
Lee, F. L. (2006). Cultural discount and cross-culture predictability: Examining the box office performance of American movies in Hong Kong. *Journal of Media Economics*, *19*(4), 259–278. doi:10.1207/s15327736me1904_3
Lee, H. (2006). Peripherals encounter: The Hong Kong film syndrome in South Korea. *Discourse*, *28*(2/3), 98–113.
Leong, A. M. W., Yeh, S., & Chang, L. (2015a). Nostalgia tourism in Macau: The bidirectional causal relationship between destination image and experiential value. *Journal of Hospitality and Tourism Technology*, *6*(1), 89–99. doi:10.1108/JHTT-02-2015-0010
Leong, A. M. W., Yeh, S., Hsiao, Y., & Huan, T. (2015b). Nostalgia as travel motivation and its impact on tourists' loyalty. *Journal of Business Research*, *68*(1), 81–86. doi:10.1016/j.jbusres.2014.05.003
Liou, D. (2010). Beyond Tokyo Rainbow bridge: Destination images portrayed in Japanese drama affect Taiwanese tourists' perceptions. *Journal of Vacation Marketing*, *16*(1), 5–15. doi:10.1177/1356766709356137
Macionis, N. (2004). Understanding the film-induced tourist. In W. Frost, G. Croy, & S. Beeton (Eds.), *International tourism media conference proceedings* (pp. 86–97). Berwick: Monash University Tourism Research Unit.
Macionis, N., & Sparks, B. (2009). Film-induced tourism: An incidental experience. *Tourism Review International*, *13*(2), 93–101. doi:10.3727/154427209789604598
Marchegiani, C., & Phau, I. (2013). Development and validation of the personal nostalgia scale. *Journal of Marketing Communications*, *19*(1), 22–43. doi:10.1080/13527266.2010.542078
Mercille, J. (2005). Media effects on image: The case of Tibet. *Annals of Tourism Research*, *32*(4), 1039–1055. doi:10.1016/j.annals.2005.02.001
Nunnally, J. (1978). *Psychometric theory* (2nd ed.). New York, NY: McGraw-Hill.
Pan, S., & Ryan, C. (2013). Film-induced heritage site conservation: The case of Echoes of the Rainbow. *Journal of Hospitality and Tourism Research*, *37*(1), 125–150. doi:10.1177/1096348011425497
Pretes, M. (1995). Postmodern tourism: The Santa Claus industry. *Annals of Tourism Research*, *22*, 1–15. doi:10.1016/0160-7383(94)00026-O
Redondo, I. (2012). Assessing the appropriateness of movies as vehicles for promoting tourist destination. *Journal of Travel and Tourism Marketing*, *29*, 714–729. doi:10.1080/10548408.2012.720156
Reijnders, S. (2009). Watching the detectives: Inside the guilty landscapes of inspectors Morse, Baantjer and Wallander. *European Journal of Communication*, *24*(2), 165–181. doi:10.1177/0267323108101830
Reijnders, S. L. (2016). Stories that move: Fiction, imagination, tourism. *European Journal of Cultural Studies*, 19 (6), 672–689. doi:10.1177/1367549415597922
Riley, R. W., Baker, D., & Van Doren, C. (1998). Movie induced tourism. *Annals of Tourism Research*, *25*(4), 919–935. doi:10.1016/S0160-7383(98)00045-0
Riley, R. W., & Van Doren, C. (1992). Movies as tourism promotion. *Tourism Management*, *13*, 267–274. doi:10.1016/0261-5177(92)90098-R
Rittichainuwat, B., & Rattanaphinanchai, S. (2015). Applying a mixed method of quantitative and qualitative design in explaining the travel motivation of film tourists in visiting a film-shooting destination. *Tourism Management*, *46*, 136–147. doi:10.1016/j.tourman.2014.06.005
Robinson, P. (2015). I remember it well: Epiphanies, nostalgia, and urban exploration as meditators of tourist memory. *Tourism, Culture and Communication*, *15*, 87–101. doi:10.3727/109830415X14401707765881
Roesch, S. (2009). *The experiences of film location tourists*. Bristol: Channel View Publications.
Schofield, P. (1996). Cinematographic images of a city. *Tourism Management*, *17*(5), 333–340. doi:10.1016/0261-5177(96)00033-7
Sellicka, M. C. (2004). Discovery, connection, nostalgia. *Journal of Travel and Tourism Marketing*, *17*(1), 55–71. doi:10.1300/J073v17n01_04
Soliman, D. (2011). Exploring the role of film in promoting domestic tourism: A case study of Al Fayoum, Egypt. *Journal of Vacation Marketing*, *17*(3), 225–235. doi:10.1177/1356766711409183
Su, H. J., Huang, Y.-A., Brodowsky, G., & Kim, H. J. (2011). The impact of product placement on TV-induced tourism: Korean TV dramas and Taiwanese viewers. *Tourism Management*, *32*(4), 805–814. doi:10.1016/j.tourman.2010.06.020
Suni, J., & Komppula, R. (2012). SF-filmvillage as a movie tourism destination: A case study of movie tourist push motivations. *Journal of Travel & Tourism Marketing*, *29*, 460–471. doi:10.1080/10548408.2012.691397
Tooke, N., & Baker, M. (1996). Seeing is believing: The effect of film on visitor numbers to screened locations. *Tourism Management*, *17*(2), 87–94. doi:10.1016/0261-5177(95)00111-5
Tuan, Y. F. (1980). Rootedness versus sense of place. *Landscape*, *24*, 3–8.
Vesey, C., & Dimanche, F. (2003). From Storyville to Bourbon street: Vice, nostalgia and tourism. *Journal of Tourism and Cultural Change*, *1*(1), 54–70. doi:10.1080/14766820308668159

Wang, M., Li, W., & Chou, M. (2014). Nostalgia, perceived value, satisfaction, and loyalty of cruise travel. *International Journal of Organizational Innovation*, *6*(4), 184–191.

Weng, X. (2008). The decline of Hong Kong film industry. Retrieved July 12, 2016, from http://jmsc.hku.hk/hkstories/content/view/688/8530/1/0/

Wong, C., & McDonogh, G. (2001). Consuming cinema: Reflections on movies and market-places in contemporary Hong Kong. In G. Mathews & T. Lui (Eds.), *Consuming Hong Kong* (pp. 81–116). Hong Kong: Hong Kong University Press.

Wray, M., & Croy, W. G. (2015). Film tourism: Integrated strategic tourism and regional economic development planning. *Tourism Analysis*, *20*(3), 313–326. doi:10.3727/108354215X14356694891898

Yeh, S., Chen, C., & Yeh, Y. (2012). Nostalgic emotion, experiential value, destination image, and place attachment of cultural tourists. *Advances in Hospitality and Leisure*, *8*, 167–187.

Yen, C. H., & Croy, G. W. (2016). Film tourism: Celebrity involvement, celebrity worship, and destination image. *Current Issues in Tourism*, *19*(10), 1027–1044. http://dx.doi.org/10.1080/13683500.2013.816270.

Immersion in film tourist experiences

Yannik St-James, Jessica Darveau and Josyane Fortin

ABSTRACT
This study adopts appropriation theory to examine the role of film in tourist experiences. Specifically, it analyzes how tourists draw on and manipulate movies and TV programs to construct immersive experiences. Depth interviews reveal how consumers engage in nesting, investigating, and stamping operations in order to reduce the physical, mental, and emotional distance with places, plots, and characters. Findings expand the understanding of the relationship between film and tourist experiences by documenting instances of accidental film tourism wherein film plays an impromptu, meaningful role in tourist experiences, and by showing how film reconsumption extends tourist experiences beyond the trip itself.

Introduction

A fast-growing stream of interdisciplinary research explores the relationship between film and tourism, a phenomenon most commonly known as film tourism (Connell, 2012). Film tourism refers to "visitations to sites where movies and TV programmes have been filmed as well as tours to production studios, including film-related theme parks" (Beeton, 2005, p. 11) and involves, more broadly, "tourist experiences connected with films [including] movies, television and other screened media representations" (Yen & Croy, 2016, p. 1029). Film tourism researchers have sought to define the conceptual boundaries of this phenomenon, to measure its importance (Riley, Baker, & Van Doren, 1998), to assess its impact on host destinations (Connell, 2005), and to generally understand how movies and TV programs set in a particular region influence consumers' awareness and perception of this destination, as well as their motivation to visit it (Hahm & Wang, 2011; Im & Chon, 2008; Pan & Tsang, 2014).

Tourist experience has emerged as a significant area of study in this literature, as well as in tourism research more broadly. Indeed, tourism is an experiential product (Holbrook & Hirschman, 1982) and tourist experience is considered to be a driving source of value creation in the tourism industry (Volo, 2009). Pizam (2010, p. 343) even contends that "creating memorable experiences is the essence and raison d'être of the hospitality industry". A substantial body of research thus examines the nature of tourist experience (Bond, Packer, & Ballantyne, 2015; Volo, 2009), and seeks to understand how film tourists "perceive, interact, and relate to the destination" (Connell & Meyer, 2009, p. 196). In a critical review of the film tourism literature, Connell (2012) concludes that film tourist experience still constitutes an emerging area of study that warrants further research, notably with regard to the experience economy. In their work on the experience economy, Pine and Gilmore (1999) contend that immersion positively contributes to memorable experiences; this occurs when distance is abolished and the consumer is drawn into the experience. Extant research empirically demonstrates that immersion enhances the memorability of tourist experiences (Chandralal & Valenzuela, 2015). Studies on audience involvement have offered considerable insight into the role of distance and connectedness in film tourist experiences by analyzing how consumers' active film viewing experience may create a sense of connection (Kim, 2012) that motivates them to visit film locations (Fu, Haobin Ye, & Xiang, 2016) and colors their experiences as film tourists (Kim, 2012; Kim & Assaker, 2014). Yet the mechanisms underlying immersion in the tourist experience at the destination remain largely unexplored and we argue that examining the practices in which film tourists engage to co-create immersive experiences represents a useful contribution in this area.

The present study also expands our understanding of tourist experiences connected with film by considering both purposeful and incidental tourist activity (Croy & Heitmann, 2011). Indeed, whereas extant literature largely focuses on film-*induced* tourism by stressing the purposeful role of film in destination decision and tourist experiences (Connell, 2012), researchers have proposed that incidental or serendipitous film tourism represents a significant phenomenon (Croy & Buchmann, 2009; Croy & Heitmann, 2011; Macionis & Sparks, 2009)

In sum, we seek to contribute to the understanding of the role of film in tourist experiences by examining how tourists draw on and manipulate movies and TV programs to construct immersive experiences. To understand how this phenomenon plays out across both purposeful and incidental tourist experiences, this investigation is conducted in the context of leisure tourism in general. We use appropriation theory as a lens to investigate immersion in film tourist experiences.

Literature review

Film tourist experiences

In their pioneering work on the experience economy, Pine and Gilmore (1999) argue that successful experiences are memorable ones, a position that is embraced in the tourism literature (Couldry, 1998; Kim, 2010; Larsen, 2007; Pizam, 2010). Pine and Gilmore (1999) further propose that memorable experiences arise from *immersion*, that is, when distance is eliminated and the consumer becomes deeply involved in the experience. In contrast, *absorption* implies a certain distance to the experience, a situation the authors illustrate with the example of watching a movie. This is an intriguing distinction from the viewpoint of film tourist experiences as it suggests the potential for tourism to transform an absorptive experience (film consumption) into an immersive one by reducing the distance between the consumer and the film.

Consistent with this idea, extant research empirically demonstrates how immersion and proximity contribute to memorable tourist experiences. For example, Chandralal and Valenzuela's (2015) Memorable Tourism Experiences scale contains items such as "I could immerse myself in local festivals and other cultural ceremonies" and "I closely experienced the actual local cultures". Past studies document several other instances of film tourists' desire for greater proximity with a film and a destination – their wish and eagerness to "come closer" to a story, to "make a connection" (Reijnders, 2011), to become part of a "cultural landscape" (Jewell & McKinnon, 2008), or to "immerse themselves in this mythical world" (Buchmann, Moore, & Fisher, 2010). To date, the influence of distance and connectedness on film tourist experiences has been studied most fruitfully from the perspective of audience involvement. Audience involvement is a multidimensional concept that originates from media studies and construes the audience's active viewing experience as composed of emotional, cognitive, behavioral, critical, and referential dimensions (Kim, 2012; Kim & Assaker, 2014). This conceptualization encompasses and builds on the notion of parasocial interaction, which refers to the imaginary sense of intimacy an individual audience member develops with a media figure (Horton & Wohl, 1956). Extant research on audience involvement thus suggests that viewing experiences of media programs may lead to a sense of connection and intimacy with characters, stories, and the places portrayed (Kim, 2010, 2012); in turn, audience involvement impacts destination image (Yen & Croy, 2016), positively influences intentions to travel to portrayed filmed locations (Fu et al., 2016) and affects film tourist experiences (Kim, 2010, 2012; Kim & Assaker, 2014; Roesch, 2009). For instance, audience involvement has been positively associated with a propensity to immerse oneself in a film location through re-enactment of scenes from the film (Kim, 2010; Roesch, 2009) and the desire to meet a favorite character (Kim, 2012). This stream of research therefore sheds light on the audience's immersion in the film experience, and documents the impact that the perceived closeness thus developed has on film tourist experiences. In contrast, it leaves unanswered the question of how perceived distance is reduced at the destination to encourage immersion in the film tourist experience. We seek to address this issue by exploring the practices film tourists adopt to co-create immersive experiences.

A second opportunity to expand our understanding of tourist experiences connected with film is to examine experiences that occur beyond the narrow boundaries of this phenomenon in extant literature. Indeed, this research is largely conducted within the confines of film-*induced* tourism, an approach that emphasizes how film viewing motivates destination choice and tourist experiences (Connell, 2012). Previous studies indicate that purposeful film tourism represents a small portion of tourism activity related to film, which is largely an incidental phenomenon (Croy & Heitmann, 2011; Macionis & Sparks, 2009). Extant research also suggests that purposeful and incidental film tourists may experience a destination differently (Buchmann et al., 2010; Roesch, 2009). We propose that simultaneously considering both the purposeful and incidental tourist activity is a meaningful avenue from which to cultivate a richer understanding of the relationship between film and tourist experiences.

Appropriation theory and tourist experiences

Carù and Cova (2003, 2005, 2006) propose that the concept of appropriation is useful in understanding immersion in consumer experiences. Drawing from research in the field of environmental psychology, they argue that the goal of appropriation is to make something "one's own" by reducing or eliminating the distance – physical, mental, or emotional – between the person and a product, service, or space. Appropriation involves three operations that facilitate immersion: nesting, investigating, and stamping. Through *nesting*, the consumer approaches an experience by isolating a familiar element; this operation thus creates a comfortable "nest" from which the consumer can safely explore other aspects of the experience. This exploration occurs through *investigating*, as the consumer progressively considers other, less familiar products or activities in order to enrich his or her references. Ultimately, the consumer engages in *stamping* by creatively assigning a unique meaning to his or her experience; this meaning is idiosyncratic and subjective as it evolves from the consumer's personal story. Appropriation theory thus considers the consumer as co-creator of the consumption experience and appropriation operations constitute practices that allow the consumers to become immersed in this experience.

Carù and Cova (2003, 2005, 2006) empirically examined appropriation operations – and the service elements that facilitate or impede them – in the context of artistic products, more specifically classical music concerts. Earlier research in this area was grounded in appropriation of physical spaces, such as workspace (Fischer, 1983) and servicescapes (Aubert-Gamet, 1997). The notion of distance in appropriation theory can thus be construed as either physical, mental, or emotional. In this sense, appropriation theory constitutes a useful framework to cultivate our understanding of film tourist experiences since immersion in this context may simultaneously involve reducing the physical, mental, and emotional distance to the location where a movie or TV program was filmed and to the film's story and characters.

Methodology

We adopted a qualitative approach underpinned by hermeneutic phenomenology, a philosophical perspective that examines lived experience formed through meanings, understandings, and interpretations. This perspective is based on the work of several philosophers who theorized about lived experience (e.g. Heidegger, Merleau-Ponty) and narrative unfolding (e.g. Gadamer, Ricoeur). In tourism research, Pernecky and Jamal (2010, p. 1063) argue that hermeneutic phenomenology "provides researchers with the opportunity to explore how tourists' meaningful experiences come about".

Data collection involved conducting in-depth interviews (McCracken, 1988) to elicit rich descriptions of tourists' co-construction practices and of the meanings they attach to these practices (Pernecky & Jamal, 2010; Thompson, 1997). Since our goal was to examine how consumers use film in their overall tourist experiences, we sought descriptions of a variety of experiences in different destinations and we were careful to avoid imposing the theme of film tourism a priori in our interviews. This had repercussions on the selection of informants and data collection. To enhance the odds that informants had experienced film tourism, we selected travelers who had made a trip to Spain or to New York City, United States (US); both destinations have been depicted in popular movies and New York is featured in numerous TV series. Since informants lived in a large metropolitan area in the province of Quebec (Canada), Spain constituted a more faraway destination, whereas New York was a more accessible one. Second, we presented the subject of our study as an investigation of tourist experiences without mentioning explicitly the concept of film tourism. We employed an interview protocol that first profiled informants' travel history and then explored in more detail specific travel experiences. Our interview protocol was designed to allow participants to recount their travel experiences in their own terms. It involved a series of "grand-tour" questions followed by more specific prompts and follow-up questions intended to elicit a more complete description of the participants' experiences and to explore all relevant facets of those experiences (McCracken, 1988). For instance, we asked participants: "Tell me about your trip to Spain/New York City."; we then followed up with questions such as "Did you travel alone or with someone?", "How did you decide to take this trip?", "Can you tell me about a memorable experience during this trip?". When informants evoked a movie or TV program, we focused the interview on this experience and used probing questions to further explore how film participated in their tourist experience. When no film was mentioned in informants' travel stories, we then introduced the subject directly.

Informants were recruited through a snowballing technique within the researchers' personal networks: we asked acquaintances to identify people who had traveled to Spain or New York City and to share with them an invitation to participate in our study; these persons were also asked to refer other potential participants. Informants were recruited until theoretical saturation was reached (Spiggle, 1994), and a total of 13 individuals were interviewed between February and August 2010. One informant was interviewed twice since he was planning an imminent trip at the time of our first interview and interviewing him immediately before and after this trip offered a useful opportunity to examine the imagined and lived tourist experience. Interviews averaged 60 minutes and ranged from 40 to 120 minutes. Ten informants mentioned at least one movie or TV program without prompting when describing their travel experiences. Among the three persons who did not do so, two provided meaningful descriptions of film tourist experiences when prompted to do so, and only one reported no such experience. The prevalence of film-related themes in our informants' tourist experiences supports Croy and Heitmann's (2011) proposition that the relationship between film and tourism is very significant, yet complex and often subtle. A detailed profile of informants is presented in Table 1; all informant names are pseudonyms.

The interpretation of our qualitative data set was guided by Thompson's (1997) hermeneutically grounded framework for interpreting consumer stories, in this case tourist stories. Our interpretation began as we translated the interviews from French into English; as the authors are fluent in both languages, this exercise provided us with an opportunity to reflect on and discuss the meanings in informants' narratives (Temple & Young, 2004). Throughout the analysis, we considered both French and English texts to refine our understanding of participants' experiences. In order to represent more faithfully participants' meaning and context, the translations were kept as close as possible to the original French style while ensuring their readability (Dion, Sabri, & Guillard, 2014). The interpretation then involved an individual analysis to reconstruct each informant's travel history and film tourist experiences; an emphasis was placed on situating these experiences

Table 1. Informants' profile.

Pseudonym	Age (years)	Occupation	Focal film tourist experience(s)	Films discussed
Abby	45	Journalist	Stockholm (Sweden)	Millenium
Bradley	52	Actor; teacher; clerk in video store	Madrid (Spain) Barcelona (Spain)	Bienvenido a Casa L'Auberge Espagnole Jeune Perez Vicky Cristina Barcelona
Christine	43	Entrepreneur	New York (USA)	Sex and the City When Harry Met Sally Sleepless in Seattle Miami Vice Desperate Housewives
Ella	38	Secretary	New York (USA)	Sex and the City The Sopranos Lipstick Jungle 30 Rock When Harry Met Sally
Ethan	18	Student	Spain	L'Auberge Espagnole Vantage Point
Julie	24	Consultant	Barcelona (Spain)	L'Auberge Espagnole Hostel Vicky Cristina Barcelona
Lara	23	Government employee	Europe: France, Spain, England (GB), Switzerland New York (USA)	L'Auberge Espagnole Vicky Cristina Barcelona Pedro Almodóvar movies in general Pan's Labyrinth Sex and the City
Marie	32	Business development	European cruise and Barcelona (Spain)	
Patrick	42	Media librarian	Barcelona (Spain)	L'Auberge Espagnole Vicky Cristina Barcelona À la di Stasio
Simon	26	Student	Barcelona (Spain)	Vicky Cristina Barcelona Pedro Almodóvar movies in general
Sophie	24	New media coordinator	Barcelona and Madrid (Spain)	L'Auberge Espagnole Vicky Cristina Barcelona
Tina	59	Researcher	France	Le Fabuleux Destin d'Amélie Poulain Harry Potter series
Zoe	23	Dancer	New York (USA)	How I Met Your Mother Sex and the City Gossip Girl The Single Guy Remember Me

in the informant's personal history as well as on identifying the meanings informants ascribed to them. Next, a cross-case analysis was performed with the goal of identifying patterns across different informants' experiences. In these analyses, appropriation theory (Carù & Cova, 2003, 2005, 2006) was used as a "sensitizing device" (Klein & Myers, 1999, p. 75) to arrive at a first etic categorization of data. This process, however, also allowed for emic concepts and themes to emerge as we identified idiographic elements in the participants' narratives. The final stage of our analysis involved further developing our understanding of emerging themes in informants' experiences in relation to extant literature. Our hermeneutic interpretative process involved a series of part-to-whole iterations, moving between individual stories and the emerging of the complete set of textual data, with the goal of removing contradictions from our interpretation (Thompson, 1997).

Findings

This study examines the role of film in tourist experiences. More specifically, we analyze tourists' stories to understand how they draw on and manipulate movies and TV programs to construct immersive experiences. The first theme – *nesting* – describes how film conveys destination images that act as useful anchorage points when tourists first encounter that destination. The second theme fleshes out *investigating* operations consumers engage in to explore new dimensions of a destination or film. *Stamping* constitutes our third theme and it discusses how consumers creatively assign idiosyncratic meanings to their tourist and film experiences.

Nesting: recognizing the imagined site

Our informants' narratives document how exposure to various screen media shapes their mental image of a destination and, in some instances, their desire to visit this destination. When recounting why she decided to visit Spain, Sophie spontaneously mentions a movie: "*L'Auberge Espagnole* is one of my favorite movies. Barcelona and everything in it ... I adore that [...] It's an ongoing party, a really relaxed lifestyle, you eat late, you party and everything. It's sunny and warm." Sophie's narrative illustrates how the general ambiance and lifestyle depicted in a movie can constitute a strong motive in the decision to travel to a given destination (Josiam et al., 2015).

Like Sophie, Patrick's perception of Barcelona is intrinsically tied to the movie *L'Auberge Espagnole*: "We had seen *L'Auberge Espagnole* a long time ago and we wanted to watch it again before going but we didn't. When I thought of Barcelona I thought of that movie." However, Patrick also used a variety of screen media – including a TV cooking show and social media – in preparing for his trip: "I'm really looking forward to visiting two markets in Barcelona [...] I saw them last year on TV, on *A la di Stasio*. They had an episode on Spain. I also did some reading and then I looked things up on the Internet. I saw videos on YouTube." Similarly, Lara alludes to various movies when recounting her trip to Barcelona – *L'Auberge Espagnole, Pan's Labyrinth, Vicky Cristina Barcelona*, as well as Pedro Almodóvar movies in general – and yet her desire to visit the Sagrada Familia arose from watching the 1992 Summer Olympics set in Barcelona:

> [Canadian diver Alexandre Despatie] won in Barcelona when I was a kid [...] That image was crazy because he was on the springboard and you could see the city behind him. You could see that huge church. And I wanted to go see that!

Our informants also combine literary and screen media in preparing for a trip. For instance, Abby traveled to Stockholm after reading the first two books and watching the first movie in the *Millenium* trilogy, something she frequently does before a trip: "I read books before leaving, and I read touristic guides as well. I watch movies, too."

Hence, tourists become familiar with various aspects of a location by fluidly weaving together literary and screen media, including different types of screen media – such as movies, TV series, non-fiction TV shows, and social media. These practices are consistent with convergence (Månsson, 2011) and with the notion of "circles of representation" (Jenkins, 2003) wherein images projected in the media shape individual perceptions of a destination and their subsequent encounter with this destination. The familiarity thus acquired contributes to creating an experience of "really being there" when the consumer arrives at the destination. Our informants relate meaningful instances when their encounter with a site is coherent with the mental image they had formed based on a screen product. This is evidenced in Abby's description of her arrival in Stockholm and Sophie's recollection of first encounter with different destinations:

> When I got out of the taxi to reach my hotel I realized I was in Stureplan, which is one of the first scenes in the movie. That was really a wild feeling, like, hey I'm really here. It's a large square with big advertising screens and in the movie Mikeal Blomkvist runs out of the courthouse into this place after being found guilty and he's everywhere on the news [...] And so this square with all

> the big screens was just beside my hotel and I thought: "I'm really here", and I was super happy. (Abby)

> When we arrived at the spot with the benches, the spot where they kiss in *L'Auberge Espagnole*, that's a striking image. And I thought: "Ah, I saw this in one of my favorite movies. And I'm here." It was the same thing when I saw the Eiffel Tower for the first time. In London too. You see things in movies, like those two-story buses and that's fun to see. In Russia, you see the Red Square on the news and when you're there, in person, it's impressive. (Sophie)

These excerpts illustrate how exposure to screen media establishes certain objects and sites as familiar anchorage points that facilitate the nesting process by creating the impression of "being there". Indeed, Abby reports that experience as she finds herself on the site of a memorable scene in the *Millenium* and Sophie even compares a particular area in Park Güell depicted in *L'Auberge Espagnole* to iconic landmarks such as the Eiffel Tower.

Tina's story further illustrates how literary and screen media create a mental image that shapes entry into a subsequent experience as she recounts preparing for her journey "in Marie-Antoinette's footsteps":

> The Conciergerie affected me the most. I had prepared it for two days. I had read about Marie-Antoinette's last days before going there. I had really prepared my Conciergerie. So when I arrived it was a little strange because I felt like I knew the place [...] It was similar with *Millenium*. I had imagined the landscape and when I saw the movie it was exactly like that [...] I was delighted to see it was as I imagined.

Tina's narrative not only illustrates how literary and screen media both shape expectations of a destination, but it also reveals comparable nesting operations between media experience and tourist experience. Indeed, Tina reports similar experiences of recognition after reading a book and visiting a site ("I felt like I knew the place") as she does after watching a movie based on a book ("it was as I imagined").

Past studies indicate that film plays a significant part in shaping tourists' expectations of a destination and their desire to visit certain sites (Josiam et al., 2015; Kim, 2010; Pan & Tsang, 2014). Our findings further suggest that an important first component enabling immersion in a film tourist experience involves nesting operations wherein the tourist's encounter with a site is compared with a mental image shaped through screen media consumption. Kim (2010, p. 61) describes tourism as "a visual encounter with a place that is coded as distant, both spatially and temporally". From an appropriation perspective, the images portrayed in screen media create familiar anchorage points in consumers' minds that help to reduce distance when encountering these objects or places; as a result, tourists express feelings of "really being there". From this comfortable "nest" of the familiar, tourists are confident to explore additional aspects of the destination, the plot, and the characters. We examine these "investigating" operations next.

Investigating: exploring the destination and the film

From the physical encounter with an imagined site in the nesting phase, our data reveal that film tourists engage in two forms of investigating operations as they explore new activities and unfamiliar dimensions of the experience (Carù & Cova, 2003): investigating the plot and characters on-site, and revisiting the film.

Investigating the plot and characters on-site

Informants recount numerous instances where being on-site allowed them to deepen a connection they have to a film's plot or characters. For instance, Christine and Ella traveled to New York to celebrate Christine's birthday and taking a *Sex and the City* tour was an important component of this trip. The two friends are big fans of the show: They have watched the TV series and movies several times, have memorized many dialogues, and can describe what the characters wore in different scenes. Christine and Ella clearly sought an immersive experience: "We wanted to go there and to feel like we were in the show a little" (Christine). Christine recalls one such instance:

> When we had a drink at the bar that belongs to Miranda's boyfriend. It's a little weird because it's like you're walking into the studio. You get a bit closer to that which you're of fan of. They also create an ambiance [...] In the bus they have a TV and they explain: "We are arriving at such place. Remember what happened there." And they play that scene before you arrive. I mean, a TV show is not real but when you visit a site, it becomes real. That's how you feel. When it plays on TV, you're into it, you watch it, but when it's over, It's over. But then you walk into it and it takes form, it becomes real [...]

> When you're a fan [...] you feel like you are walking in their steps so you are closer to them. You are where they really are, where they've been.

Christine's desire to reduce the perceived distance ("get closer") to characters is consistent with studies documenting the importance of audience involvement in film tourism (Kim, 2010, 2012). Her perception that the story "becomes real" through this visit further recalls the concept of *lieux d'imagination* as "physical points of reference, such as objects or places, which, for specific groups in the society, provide the opportunity to construct and subsequently cross the symbolic boundary between an 'imagined' and a 'real' world" (Reijnders, 2010, p. 40). Christine's story also illustrates the central role of the tour operator in facilitating distance reduction in the *lieux d'imagination*. In their analysis of investigating operations in the context of classical concerts, Carù and Cova (2005, 2006) insist on the importance of the Maestro as a guide. The tour operator plays a similar role here, creating an "ambiance" and providing participants with the required knowledge to investigate a site.

Abby also wishes to become closer to characters through film tourism: "I want to get closer to the characters, to be there with them, to identify completely with them and to live experiences with them, in the same places." She further recounts traveling to Stockholm and creating her own itinerary in the footsteps of *Millenium*'s character Lisbeth Salander:

Abby : I went to the gorgeous apartment Lisbeth Salander buys when she becomes rich. I went to the café where she goes at one point, to the bar where she goes at the end. I even went to the place where she buys her groceries. She eats some horrible stuff, like frozen pizza. I didn't eat that – I mean, there are limits – but I went to the same café to buy similar food. I'm a total Lisbeth Salander fan. I am!

Interviewer : And how did you feel going to that grocery store?

Abby : I was hoping to see her. I was hoping to see signs of her, some references. I was a little disappointed that in that grocery store, everyone acts like they don't know *Millenium*. There is no tribute to Lisbeth Salander. Nothing. [...] The city could have done more to make me live an experience, but again no because then you feel it's staged and it's not real anymore.

Like Christine and Ella, Abby engages in re-enactment in order to get closer to a character (Kim, 2010), going to "the same café to buy similar foods". However, Abby expresses disappointment that there was no "tribute" to the *Millenium* character, claiming she was "hoping to see signs of [Lisbeth Salander], some references". This resonates with Croy and Buchmann's (2009, p. 154) observation that "[t]ourists will come, demanding experiences informed in part by films" and that these expectations compete with traditional destination histories. The absence of references to guide her investigating thus hinders Abby's efforts to reduce the distance with the character. Yet Abby reflects on the fact that "staged" references might lessen the authenticity of her experience ("it's not real anymore").

Investigating operations can also be impeded by environmental factors that transform the site into a less credible stage for re-enactment. Consider how Abby describes finding it more difficult to be immersed in her experience ("to truly soak in it") because of the light:

> *Millenium* is set in a very dark atmosphere and when I visited Stockholm it was spring so it was sunny and the days were longer [...] I wasn't unhappy that the weather was nice but I knew that it didn't fit [...] I would have had to go there in January to truly soak in it.

In sum, our data suggest that film tourists become more immersed in an experience as they engage in investigating operations by exploring new areas of the destination and re-enacting certain scenes. Immersion occurs as they thus reduce the distance to characters with whom they have formed an attachment and with the plot (Kim, 2010). Service providers and servicescapes can facilitate this process by providing an appropriate ambiance and useful references.

Revisiting the film

Our informants' investigating operations are not limited to on-site exploration and, interestingly, they describe revisiting a film after their trip. This practice is a form of reconsumption, that is, "consumption experiences that consumers actively and consciously seek to experience again" (Russell & Levy, 2012, p. 341). In these instances, informants note how they have acquired a certain familiarity with locations and cultural practices, which then allows them to investigate the film with enhanced knowledge of the context in which it takes place (Carù & Cova, 2003). We previously noted how *L'Auberge Espagnole* inspired Sophie's trip to Barcelona; she also recounts watching the movie again after her trip:

> I watched it again. It was fun. You see places where you've been. And you sometimes see things differently than when you were really there. When you watch the movie, you think that a certain place is in a certain location in the city. But [having been there] you know better where it is geographically. I also like this movie because I was on a student exchange program so I know that feeling [...] I have had that experience and that's always something you wish you could do again but that part of your life is over. So the fact that this is

> an experience I've had in a city where I've been to ... there are many elements I can relate to.

Sophie uses the knowledge of Barcelona she acquired during her trip to gain a greater understanding of the movie's context upon reconsumption. She also understands the plot at a personal level since she has been on a student exchange program. This investigating process reveals a richer relationship between film and tourism than previously portrayed in the film-induced tourism literature. Indeed, not only is Sophie's tourist experience motivated and shaped by a movie, it in turn colors her subsequent reconsumption of that movie by enabling further investigating. Patrick clearly articulates this relationship when he discusses wanting to watch *L'Auberge Espagnole* in an interview immediately following his trip to Spain:

> I talked about watching it again after coming back. We need to watch it again. Now we could watch the movie and make some associations. We would know where the movie took place. I would better understand the geography. When I watched it the first time I didn't know anything about the city. I mean nothing. So I thought it was nice but I didn't know where it was. No reference at all. Now I would know!

Sophie and Patrick both discuss how *L'Auberge Espagnole* played a role in their tourist experience. In contrast, film does not play an explicit role in Ethan's tourist experience and yet his trip to Spain transforms his subsequent reconsumption of the movie *Vantage Point*:

> I had watched it once but I didn't think about it when I went to Spain. I saw it sometime after coming back and I thought: "Wow! I saw this before and I didn't even think about it when I was in Spain." It reminded me of good memories. I didn't visit the spot where the main scene in the movie takes place but everything else made me think of Spain [...] the streets crowded with people speaking Spanish, summer and the warm weather, the little taxis, all those little things that remind you of Barcelona.

Tourists' on-site experiences also appear to transform engagement in film reconsumption; consider Lara's story:

> I have been to New York before and after watching *Sex and the City*. It really shows New York; you can feel New York. And it's fun to discover ... "Oh, they're in the Upper East Side; Oh, now they're in Soho, now they're in Chinatown, in the Financial District. Oh, they went to the same restaurant I went to." You know, there's a scene in the movie where they go to a restaurant and I had been there before they made the movie. That's cool. When I saw the movie and I had been there before them I was, like "Hey, you're late!" [*laughs*] But it reminds you of good memories. It makes you want to go back.

Because of her familiarity with New York, Lara has a clearer understanding of the sites featured in *Sex and the City*. Interestingly, having been to a restaurant prior to watching the movie brings her closer to the plot and characters as she symbolically interacts with them ("Hey, you're late!"). Lara's narrative illustrates how investigating operations through reconsumption of a film can constitute a form of symbolic reconsumption of a tourist experience ("it reminds you of good memories") as well as a motivation to travel to the destination again ("It makes you want to go back").

Traveling to a film location broadens tourists' understanding of the context and allows for investigating operations through film reconsumption. All our informants mentioned having watched a film again – or wanting to watch it again – after making a trip to a location it depicted. This meaningful phenomenon illustrates that the film tourist experience does not end upon returning from a trip and suggests that it also entails an important symbolic component that occurs afterward.

Stamping: meaning-making and the film tourist experience

Stamping involves creatively and subjectively imbuing an experience or a portion of an experience with unique meaning (Carù & Cova, 2003). Our data reveals two forms of meaning-making in the context of film tourist experiences: becoming an accidental film tourist, and creating one's own experience.

The Accidental film tourist

Our informants' tourism stories relate situations where a film plays an impromptu yet meaningful role in their tourist experience. Consider, for instance, Julie's description of an evening in Barcelona:

> One evening we went out with everyone from the youth hostel and it was just like in *L'Auberge Espagnole*! Everyone speaking different languages. We were singing in the street, trying to find songs everyone knew [...] Everyone was happy. It was like a party, it was really fun [...] My sister and I just looked at each other and we were, like: "This is just like in *L'Auberge Espagnole*!"

Julie had watched *L'Auberge Espagnole* prior to her trip but she does not describe her decision to visit Spain or to stay in a youth hostel in Barcelona as influenced by this movie. She nevertheless vividly recalls *L'Auberge Espagnole* when finding herself in a situation similar to that depicted in the movie. Likewise, Zoe reports this encounter in a New York bar:

> I was standing in line for a drink and it was packed with people. And this guy comes up to me and says: "Hey, have you met Sam?" and he swings his friend Sam in front of me. I couldn't believe it because it's the opening line in *How I Met Your Mother* and I really liked that show. It just made my evening.

Whereas Christine and Ella intentionally traveled to New York to live a *Sex and the City*-themed experience, Zoe unexpectedly finds herself in a scene that recalls a favorite show. Contrary to a commercially scripted experience, this represents a situation where three individuals (Zoe, Sam, and Sam's friend) draw on a popular TV series to shape their interaction:

> I've watched this TV show often [...] and I felt this was something I had heard often and that it was part of my life. And then this stranger gives me that line while we're in New York. I thought it was excellent, that the timing was perfect, as if we shared a bond but we didn't, really [...] And he thought it was cool that I reacted that way. He asked me if I knew the TV show and I said: "Of course!" So I spent the evening with this Sam and his friends and it was really, really fun.

Zoe and Julie's narratives evince how film can play an unplanned yet meaningful role in tourists' experiences, contributing to transform rather commonplace events – such as an evening out with friends – into memorable experiences. In these instances, tourists spontaneously draw on familiar film elements – such as plots, characters, and settings – to interpret and give meaning to their experience. We label these events "accidental film tourist experiences" to acknowledge their unplanned, fortuitous nature. They constitute an empirically derived form of experience that can be associated with Croy and Heitmann's (2011) notion of a serendipitous film tourist for whom film has low importance in the destination decision but a profound role in the destination experience. Such experiences have largely been overlooked in extant literature, a fact probably attributable to its focus on film-induced tourism, that is, tourist visits that are explicitly motivated by film (Connell, 2012).

Creating one's own experience

Our informants share stories where a film was used as a resource in interpreting a destination (nesting) but where their own knowledge and exploration of the destination (investigating) led them to reject a commercial or artistic proposition to creatively construct their own vision of the experience. For instance, Simon is an architecture student and he largely describes his trip to Spain in terms of architecture, alluding to Pedro Almódovar's movies while rejecting the vision of Spain's architecture they depict:

> In movies, like Almodovar's movies, it's often kitsch, like Gaudi's stuff. For me that's not ... Gaudi's stuff is really isolated in the city. It's not beautiful. That's not really what the city is about for me. It's just a building. I mean, they're cool buildings. They represent an era with a distinctive signature. There are thousands of small projects in Barcelona but the city is still fairly homogeneous in terms of materials and architecture and ... Gaudi just clashes with that.

Some informants engage in this creative process when their encounter with a site is disappointingly inconsistent with the mental image created through film consumption. Lara offers such an example when she discusses her visit to Park Güell in Barcelona and compares it to a scene in *L'Auberge Espagnole*:

> What you see on the screen is never like reality. They only show you some part of it. With Park Güell, for example, we all remember when he kisses the girl on the steps when the sun is rising or setting. We all remember that. But that's not the best part of the park. The most incredible part is when you go under that place and there are musicians, and you look up at the ceiling and it's still in stones, and it's even more wonderful than anything you've seen. [...] The person who wrote that film ... something must have happened for him at Park Güell and that's where he had his moment. But that's not where I had my moment because there were lots of birds, lots of noisy kids. There were people playing the trumpet right in your face that you had to pay. That's not where I had my moment but it's still a nice place.

Lara approaches Park Güell with specific references from a scene in *L'Auberge Espagnole*, but the perceived site does not live up to the imagined one. By further investigating the site, Lara ultimately rejects the artistic vision conveyed in the movie ("that's not where I had my moment") and articulates one that is more meaningful for her ("the most incredible part").

This idea of making an experience, or portion of an experience, one's own is at the core of appropriation (Carù & Cova, 2003) and it appears frequently in our informants' narratives. For instance, Lara emphasizes "my moment" at Park Güell, Tina relates how she "really prepared my Conciergerie", and Abby discusses the importance of "having my own experience": "I never took the *Sex and the City* tour in New York. When it becomes too commercial I'm just not interested. I need to feel I'm having my own experience." This stamping operation converges with the notion of existential authenticity and "mineness" in tourist experiences, that is, "a desire to make up one's own mind about what is going on during the tourist experience, to interpret this for oneself" (Steiner & Reisinger, 2006, p. 307)

Discussion and conclusions

This study analyzes how tourists use film to construct immersive experiences by engaging in appropriation operations. Through *nesting* operations, tourists approach a new destination by comparing it with a mental image formed through screen media consumption; coherence between the imagined and perceived place reduces distance and contributes to an impression of "really being there". From this comfortable nest of the familiar, tourists initiate *investigating* operations as they explore places and re-enact movie and TV scenes with the goal of "getting closer" to film plots and characters. Investigating also occurs as individuals reconsume a film following a trip to a destination it depicts, and as they use the familiarity developed through this trip to explore new facets of the film. Finally, tourists engage in *stamping* operations by creatively imbuing an experience with idiosyncratic meaning; this occurs when consumers spontaneously evoke a movie or TV series to shape and make sense of a tourist experience, and when they challenge an artistic or commercial proposition to craft an experience as "their own".

These findings advance the understanding of the relationship between film and tourist experiences in two important ways. First, it complements and extends work analyzing how audience involvement impacts film tourist experiences by shedding light on the practices tourists engage in to co-create on-site immersive experiences. For instance, our findings are consistent with studies that demonstrate the role of film in shaping tourists' expectations of a destination and their desire to visit certain sites (Kim, 2010, 2012). An appropriation perspective further suggests that images portrayed in films create familiar anchorage points in consumers' minds that help to reduce distance when encountering these places; as a result, tourists express feelings of "really being there". Similarly, our participants' stories converge with past studies documenting film tourists' re-enactment of film scenes to feel "closer" to a character or plot (Kim, 2010; Reijnders, 2011); our analysis suggests that authenticity constitutes another motivation of this investigating practice that allows film tourists to merge the distance between imagined and real worlds (Reijnders, 2010).

Second, our findings expand the boundaries of the relationship between film and tourism. In extant literature, this relationship has largely been examined in terms of film-induced tourism, focusing on film as a motivator for tourism (Connell & Meyer, 2009). By adopting a broader approach and examining how tourists draw on film in constructing immersive experiences, we inductively uncover two under-examined phenomena: accidental film tourist experiences and film reconsumption. Accidental film tourist experiences occur when film plays an impromptu role in a tourist experience. In contrast to film-induced tourism, accidental film tourism arises as tourists unexpectedly find themselves in a situation that recalls a scene from a movie or TV program, and as they draw on those references to interpret their experience. These fortuitous, "unstaged" events stand out as particularly memorable experiences in informants' stories. The concept of accidental film tourist experiences empirically extends Croy and Heitmann's (2011) notion of a serendipitous film tourist. If film can impact destination awareness, familiarity, image, and expectations in a way that may induce tourism (Croy, 2010), we show how these associations may also play a spontaneous role for accidental film tourists as they draw on them to make sense of their experiences.

Our findings also put forward film reconsumption as an important practice in film tourism. Reconsumption occurs as tourists watch a movie or TV program again after traveling to a destination it depicts. It was omnipresent in our informants' stories, suggesting that the tourist experience does not end upon returning home from a trip. Russell and Levy (2012) examined the motivations and outcomes of reconsumption practices, including movies and trips. In their analysis, these were considered as distinct hedonic experiences. Our findings provide evidence of synergistic dynamics in reconsumption: watching a movie or TV program again after a trip can represent a form of symbolic reconsumption of the tourist experience, and it can also create a motivation to actually reconsume the destination.

Finally, our findings present useful implications for destination managers by providing practical guidance to assist in film tourists' co-creation of immersive experiences. Indeed, an appropriation perspective suggests an opportunity for service providers and servicescapes to act as "guides" in providing the anchorage points necessary for film tourists' investigating operations on-site. Even participants who rejected commercial tours and scripted experiences as inauthentic sought film references on-site to guide their investigating. They also acknowledged how environmental factors, such as crowding, can facilitate or impede immersion. At the same time, our findings highlight the importance for destination managers to allow sufficient room for film tourists to make an experience their own. For instance, participants who found commercial tours valuable in investigating plots and characters on-site also described how their own undirected investigation following the tour contributed to creating a memorable experience. A way to assist film tourists' stamping operations would be to offer resources for them to expand on commercial experiences on their own, such as indicating related sites or activities that are not comprised in a tour but that might be explored later.

In conclusion, this study speaks to the meaningful role of film in tourist experiences. Indeed, most of our informants spontaneously discussed movies and TV programs when recounting tourism stories without being prompted to think in those terms. Among the three persons who did not do so, two provided rich descriptions of film tourist experiences when prompted; the only participant who could not share such an experience still expressed her intention to reconsume a certain movie after traveling to a destination it depicted. This study also proposes that appropriation theory provides a useful framework from which to cultivate the understanding of the role of film in memorable tourist experiences. It suggests the relevance of a broader conceptualization of the relationship between film and tourist experiences, proposing the concepts of accidental film tourist experiences and film reconsumption as meaningful dimensions of the film experience. Future research is needed to fully examine these phenomena.

Disclosure statement

No potential conflict of interest was reported by the authors.

References

Aubert-Gamet, V. (1997). Twisting servicescapes: Diversion of the physical environment in a re-appropriation process. *International Journal of Service Industry Management*, *8*(1), 26–41. doi:10.1108/09564239710161060

Beeton, S. (2005). *Film-induced tourism*. Clevedon: Channel View Publications.

Bond, N., Packer, J., & Ballantyne, R. (2015). Exploring visitor experiences, activities and benefits at three religious tourism sites. *International Journal of Tourism Research*, *17*, 471–481. doi:10.1002/jtr.v17.5

Buchmann, A., Moore, K., & Fisher, D. (2010). Experiencing film tourism: Authenticity & fellowship. *Annals of Tourism Research*, *37*(1), 229–248. doi:10.1016/j.annals.2009.09.005

Carù, A., & Cova, B. (2003). Approche empirique de l'immersion dans l'expérience de consommation: Les opérations d'appropriation. *Recherche Et Application Marketing*, *18*(2), 47–65. doi:10.1177/076737010301800203

Carù, A., & Cova, B. (2005). The impact of service elements on the artistic experience: The case of classical music concerts. *International Journal of Arts Management*, *7*(2), 39–54.

Carù, A., & Cova, B. (2006). How to facilitate immersion in a consumption experience: Appropriation operations and service elements. *Journal of Consumer Behaviour*, *5*, 4–14. doi:10.1002/(ISSN)1479-1838

Chandralal, L., & Valenzuela, F. R. (2015). Memorable tourism experiences: Scale development. *Contemporary Management Research*, *11*(3), 291–310. doi:10.7903/cmr.13822

Connell, J. (2005). Toddlers, tourism and Tobermory: Destination marketing issues and television-induced tourism. *Tourism Management*, *26*, 763–776. doi:10.1016/j.tourman.2004.04.010

Connell, J. (2012). Film tourism – Evolution, progress and prospects. *Tourism Management*, *33*, 1007–1029. doi:10.1016/j.tourman.2012.02.008

Connell, J., & Meyer, C. (2009). Balamory revisited: An evaluation of the screen tourism destination-tourist nexus. *Tourism Management*, *30*, 194–207. doi:10.1016/j.tourman.2008.06.001

Couldry, N. (1998). The view from inside the 'simulacrum': Visitors' tales from the set of Coronation Street. *Leisure Studies*, *17*(2), 94–107. doi:10.1080/026143698375178

Croy, W. G. (2010). Planning for film tourism: Active destination image management. *Tourism and Hospitality Planning & Development*, *7*(1), 21–30. doi:10.1080/14790530903522598

Croy, W. G., & Buchmann, A. (2009). Film-induced tourism in the high country: Recreation and tourism contest. *Tourism Review International*, *13*, 147–155. doi:10.3727/154427209789604642

Croy, W. G., & Heitmann, S. (2011). Tourism and film. In P. Robinson, S. Heitmann, & P. Dieke (Eds.), *Research themes in tourism* (pp. 188–204). Wallingford: CABI Publishing.

Dion, D., Sabri, O., & Guillard, V. (2014). Home sweet messy home: Managing symbolic pollution. *Journal of Consumer Research*, *41*, 565–589. doi:10.1086/676922

Fischer, G. N. (1983). *Le travail et son espace: De l'appropriation à l'aménagement*. Paris: Dunod.

Fu, H., Haobin Ye, B., & Xiang, J. (2016). Reality TV, audience travel intentions, and destination image. *Tourism Management*, *55*, 37–48. doi:10.1016/j.tourman.2016.01.009

Hahm, J., & Wang, Y. (2011). Film-induced tourism as a marketing vehicle for destination marketing: Is it worth the efforts? *Journal of Travel and Tourism Marketing*, *28*(2), 165–179. doi:10.1080/10548408.2011.546209

Holbrook, M. B., & Hirschman, E. C. (1982). The experiential aspects of consumption: Consumer fantasies, feelings, and fun. *Journal of Consumer Research*, *9*(September), 132–140. doi:10.1086/208906

Horton, D., & Wohl, R. R. (1956). Mass communication and parasocial interaction. *Psychiatry*, *19*, 215–229. doi:10.1080/00332747.1956.11023049

Im, H. H., & Chon, K. (2008). An exploratory study of movie-induced tourism: A case of the movie The Sound of Music and its locations in Salzburg, Austria. *Journal of Travel and Tourism Marketing*, *24*(2–3), 229–238. doi:10.1080/10548400802092866

Jenkins, O. (2003). Photography and travel brochures: The circle of representation. *Tourism Geographies*, *5*(3), 305–328. doi:10.1080/14616680309715

Jewell, B., & McKinnon, S. (2008). Movie tourism – A new form of cultural landscape? *Journal of Travel & Tourism Marketing*, *24*(2–3), 153–162. doi:10.1080/10548400802092650

Josiam, B. M., Spears, D. L., Pookulangara, S., Dutta, K., Kinley, T. R., & Duncan, J. L. (2015). Using structural equation modeling to understand the impact of Bollywood movies on destination image, tourist activity, and purchasing behavior of Indians. *Journal of Vacation Marketing*, *21*(3), 251–261. doi:10.1177/1356766715573650

Kim, S. (2010). Extraordinary experience: Re-enacting and photographing at screen tourism locations. *Tourism and Hospitality Planning & Development*, *7*(1), 59–75. doi:10.1080/14790530903522630

Kim, S. (2012). Audience involvement and film tourism experiences: Emotional places, emotional experiences. *Tourism Management, 33*(2), 387–396.

Kim, S., & Assaker, G. (2014). An empirical examination of the antecedents of film tourism experience: A structural model approach. *Journal of Travel & Tourism Marketing, 31*(2), 251–268. doi:10.1080/10548408.2014.873316

Klein, H. K., & Myers, M. D. (1999). A set of principles for conducting and evaluating interpretive field studies in information systems. *MIS Quarterly, 23*(1), 67–94. doi:10.2307/249410

Larsen, S. (2007). Aspects of a psychology of the tourist experience. *Scandinavian Journal of Hospitality and Tourism, 7*(1), 7–18. doi:10.1080/15022250701226014

Macionis, N., & Sparks, B. (2009). Film-induced tourism: An incidental experience. *Tourism Review International, 13*, 93–101. doi:10.3727/154427209789604598

Månsson, M. (2011). Mediatized tourism. *Annals of Tourism Research, 38*(4), 1634–1652. doi:10.1016/j.annals.2011.02.008

McCracken, G. (1988). *The long interview, qualitative research methods*. Newbury Park: Sage Publications.

Pan, S., & Tsang, N. (2014). Inducible or not – A telltale from two movies. *Journal of Travel & Tourism Marketing, 31*(3), 397–416. doi:10.1080/10548408.2014.883345

Pernecky, T., & Jamal, T. (2010). (Hemeneutic) phenomenology in tourism studies. *Annals of Tourism Research, 37*(4), 1055–1075. doi:10.1016/j.annals.2010.04.002

Pine, B. J., & Gilmore, J. H. (1999). *The experience economy*. Boston: Harvard Business School Press.

Pizam, A. (2010). Creating memorable experiences. *International Journal of Hospitality Management, 29*, 343. doi:10.1016/j.ijhm.2010.04.003

Reijnders, S. (2010). Places of the imagination: An ethnography of TV detective tour. *Cultural Geographies, 17*(1), 37–52. doi:10.1177/1474474009349998

Reijnders, S. (2011). Stalking the Count: Dracula, fandom and tourism. *Annals of Tourism Research, 38*(1), 231–248. doi:10.1016/j.annals.2010.08.006

Riley, R., Baker, D., & Van Doren, C. S. (1998). Movie induced tourism. *Annals of Tourism Research., 25*(4), 919–935. doi:10.1016/S0160-7383(98)00045-0

Roesch, S. (2009). *The experiences of film location tourists*. Bristol: Channel View Publications.

Russell, C. A., & Levy, S. J. (2012). The temporal and focal dynamics of volitional reconsumption: A phenomenological investigation of repeated hedonic experiences. *Journal of Consumer Research, 39*, 341–359. doi:10.1086/662996

Spiggle, S. (1994). Analysis and interpretation of qualitative data in consumer research. *Journal of Consumer Research, 21*, 491–503. doi:10.1086/jcr.1994.21.issue-3

Steiner, C. J., & Reisinger, Y. (2006). Understanding existential authenticity. *Annals of Tourism Research, 33*(2), 299–318. doi:10.1016/j.annals.2005.08.002

Temple, B., & Young, A. (2004). Qualitative research and translation dilemmas. *Qualitative Research, 4*(2), 161–178. doi:10.1177/1468794104044430

Thompson, C. J. (1997). Interpreting consumers: A hermeneutical framework for deriving marketing insights form the texts of consumers' consumption stories. *Journal of Marketing Research, 34*(4), 438–455. doi:10.2307/3151963

Volo, S. (2009). Conceptualizing experience: A tourist based approach. *Journal of Hospitality Marketing & Management, 18*(2–3), 111–126. doi:10.1080/19368620802590134

Yen, C.-H., & Croy, W. G. (2016). Film tourism: Celebrity involvement, celebrity worship and destination image. *Current Issues in Tourism, 19*(10), 1027–1044. doi:10.1080/13683500.2013.816270

PART 3

Visuals and TV travel programs

The role of visual media in religious tourists' destination image, choice, and on-site experience: the case of Tinos, Greece

Matina Terzidou, Dimitrios Stylidis and Konstantinos Terzidis

ABSTRACT
Although visual media are known to influence the image people have of tourism destinations worldwide, only a few studies have examined their role in the context of religious tourism. This study explores the role television (TV) news and documentaries play in religious tourists' destination image, choice, and on-site experience. Ethnographic research (31 participant observations and 38 in-depth interviews) was conducted in two organized coach trips to the sacred island of Tinos, Greece. The findings suggest that broadcast images of the holy icon, the leading politicians, and the performing religious crowd shape the image of Tinos and influence visiting patterns. Additionally, the visual projection of religious tourists' offerings appears to condition peoples' on-site experience. The theoretical and practical implications of the study are also discussed.

Introduction

Tourism, as a fantasy industry, largely relies on the projection of favorable destination images to entice people to visit a destination (Pan, 2011). Especially within the religious context, most religious institutions use visual media among other means to instruct people to visit sacred places and to engage in religious performances, acknowledging the power of places and the performances in them in generating particular experiences and establishing religious belief (Bourdieu, 1991). Religious tourism is commonly viewed as a physical encounter with the sacred (Eliade, 1969), which, according to Turner (1973, p. 191) can only be evidenced outside one's surroundings, namely "out there" where the principle sites of religious interest are located.

Such a religiously motivated form of tourism is estimated to be as old as religion itself (Kaelber, 2006). Religion and tourism are historically related through the institution of pilgrimage, which is not an exceptional feature of Christianity, but, according to Lanczkowski (1982), a universal phenomenon of religious history. Pilgrimages in Hindu, for instance, are considered a religious obligation during free time (Singh, 2006). In Islam, adherents are instigated to travel to circulate Allah's word, to enjoy Allah's creations, and to enrich their knowledge (Timothy & Iverson, 2006). In a similar vein, while there are no specific references to tourism in the Bible, it can be inferred that a person's behavior and experience during leisure time should lead to a greater appreciation of God (Vukonic, 2006). While previous studies focus on the extent to which modern tourism practices can be distinct from, or comparable to, traditional pilgrimage (Boorstin, 1964; Graburn, 1989; Smith, 1992), this study asserts that the performance of tourism can accumulate as many sacred features as pilgrimage can accumulate secular ones. As Turner and Turner (1978, p. 20) state: "... a tourist is half a pilgrim, if a pilgrim is half a tourist". Within this context, religious tourists, similar to other tourists, are influenced by representational means such as myths, narratives (Badone, 2007; Bell, 2003; Selwyn, 1996), and visual media (Scarles, 2009; Siripis, Scarles, & Airey, 2013) that pre-program their expectations and experiences (Kim & Assaker, 2014; Pan & Tsang, 2014; Siripis et al., 2013), creating spaces for collective gazing and performances (Baerenholdt, Haldrup, Larsen, & Urry, 2004; Urry, 1990).

Specifically, visual media – being the focus of this paper – including television (TV) news and films are part of people's daily cultural life and shape the image they have of tourism destinations worldwide (Iwashita, 2008; Kim, Kim, Agrusa, & Lee, 2012; Morgan, Pritchard, & Piggott, 2002; Pan, Santos, & Kim, 2016). News coverage and popular television programs – deeply embedded in everyday life – are able to provide substantial information about a place in a rather short period of time (Kim & Richardson, 2003). This type of information is perceived as unbiased by consumers, as they are considered more genuine, in contradistinction to induced image formation agents like traditional advertising (Gartner, 1993; Govers, Go, & Kumar, 2007). Past research on visual media and destination image has mainly focused on how films, TV commercials, and photos affect people's destination image (Gong & Tung, 2016; Pan, 2011; Pan, Lee, & Tsai, 2014; Shani, Chen, Wang, & Hua, 2010; Yen & Croy, 2016), destination choice (Hudson, Wang, & Gil, 2011; Wong & Lai, 2013), and their on-site experience and behavior (Kim, 2010; Siripis et al., 2013; Yen & Teng, 2015). Studies, in particular, have found that visual media increased people's awareness of places, had tourist-inducing effects and influenced visitors' on-site experience (Kim, 2012; Kim, Agrusa, Lee, & Chon, 2007; Kim & Assaker, 2014; Pan & Tsang, 2014; Riley, Baker, & van Doren, 1998; Siripis et al., 2013).

However, there is a scarcity of research on the role of visual media in the context of religious tourism. Understanding such phenomena is critical given the popularity of religious tourism and the fierce competition among destinations to attract potential visitors. Building upon the work of Pan (2009) and Pan et al. (2016) on visuals, Kim (2010, 2012) and Siripis et al. (2013) on destination image and behavior, and within the specific framework of Greek Orthodoxy (Andriotis, 2009, 2011), this study seeks to extend past research on religious tourism by exploring the role TV news and documentaries play in religious tourists' destination image, choice, and on-site experience along with any implications these may have for the management and marketing of the tourism industry. This research contributes to a better understanding of the destination image formation processes, especially within the religious tourism context. It also extends our understanding of the religious tourism market, and in particular the Christian Orthodox one, which has been studied less in comparison with other Christian affiliations (Andriotis, 2009, 2011). To set the theoretical framework within which religious tourists' behavior should be looked at, the study now turns to discussing the effect of visual media on destination image, selection of sites, and on-site experiences, after first providing a broader understanding of religion and its relation to visuals.

Literature review

Religion and visuals

Religion is commonly defined as a "total mode of interpreting and living of life" (Geering, 2002, p. 147) as it helps people to ascribe meaning to their experiences and surroundings, and forms their behavior (Cohen & Rankin, 2004; Eidelson & Eidelson, 2003; Geertz, 1973; Silberman, 2005). Evidence for the key role religion plays in people's daily activities is found in practices including eating and drinking (Poulson, Eppler, Satterwhite, Wuensch, & Bass, 1998), social life (Bailey & Sood, 1993), and tourism (Cohen, 1998; Rinschede, 1992) as people have been observed to adjust their everyday conduct to their religious beliefs. Especially in relation to tourism, religious affiliations are known to affect religious tourists' decision making (Poria, Butler, & Airey, 2003; Williams, Francis, Robbins, & Annis, 2007), on-site experience, and behavior (Cohen, 1998; Rinschede, 1992); a phenomenon termed "theology of tourism" by Cohen (1998). Based on this, Jews, for example, do not travel on Saturdays and other Jewish holidays (Fleischer & Pizam, 2002). Cohen-Ioannides and Ioannides (2006), moreover, discuss how Orthodox Judaism's dietary restrictions cause some Jewish travelers to frequent hotels, airlines, and restaurants that serve kosher meals. Timothy and Iverson (2006) also note that devout Muslims will frequent hotels that have gender-segregated swimming pools and prayer rooms, are located near mosques, and have restaurants that offer halal foods.

Religious scripts/doctrines, similar to theatrical scenarios, determine the nature of performative encounters in the religious tourist experience, while religious producers, such as priests, organize the religious scenario, mobilizing stages, directing believers, materializing and aesthetizing places in their attempt to perpetuate and establish particular meanings related to sacred places and performances. Visual media play a dominant role within this respect. In Christianity, the fundamental teachings along with the religious scripts that circulate in schools, religious sites, and TV programs, such as documentaries and news, convey to religious adherents the message that a visit to a sacred place will free them from their spiritual or secular anguishes (Cohen, 1998; Mattila, Apostolopoulos, Somnez, Yu, & Sasidharan, 2001). A visit to such places constitutes an essential tool and presupposition for some religious tourists to achieve particular end results from their trip including pain alleviation, curing of the body, and restored balance and harmony (Bolen, 1994). As Bourdieu (1991) emphasized, religious institutions structure the dependence on religious symbolic power and hence the longing for it. Places outline practices and religious-oriented tourists are in fact considered "cast members" (Crang, 1997), by performing a series of directed religious practices as established by religious producers. Thus, what is projected in visual media – often constructed and organized by religious authorities and destination managers – might shape to a large extent religious tourists' destination image, selection of places, and their on-site experience.

Visual media: destination image and selection of places

Considering the intangible nature of the tourism product, visual information is critical in destination marketing as it portrays the actuality of the place and serves as a pre-taste of the destination until actual visiting takes place (Fakeye & Crompton, 1991). Information provided through visual media leads to increased familiarity with the destination, which in turn affects destination image formation (Baloglu, 2001; Fakeye & Crompton, 1991). Destination image, commonly defined as the sum of beliefs, ideas, and impressions that a person has of a destination (Kotler, Haider, & Rein, 1993), is considered critical for the success of any tourist destination as it is known to influence tourists' destination selection process, on-site experience, and their future behavior (see for example Chi & Qu, 2008; Lee, Lee, & Lee, 2005; Stylidis, Belhassen, & Shani, 2015). Tourism authorities, therefore, need to develop and maintain a positive destination image to improve the competitiveness of a place and increase visitors' loyalty (Qu, Kim, & Im, 2011; Stylidis, Shani, & Belhassen, 2017).

Commonly categorized under Gartner's (1993) autonomous image-formation agents, visual media including news reports, popular culture products (i.e. television programs), and documentaries appear to be more influential on destination image formation than the induced image formation agents (i.e. paid advertising) as they tend to have higher credibility and greater market penetration (Gartner, 1993; Govers et al., 2007). For example, in a study conducted by Govers et al. (2007), the autonomous image formation agents were ranked first by respondents as key information sources while choosing a destination. Similarly, Schofield (1996) argued that people's images of places are shaped to a large extent by TV programs which do not encompass the perceptual bias of promotional material. Past studies have further revealed the positive effect of films, TV dramas, and travel documentaries on destination image and visiting patterns (Kim, Agrusa, Chon, & Cho, 2008; Kim et al., 2007; Kim & Assaker, 2014; Pan & Tsang, 2014; Riley et al., 1998). Kim and Richardson (2003), for instance, reported that a movie related to Vienna had a significant positive effect on some components of Vienna's image and on respondents' intention to visit this destination in the future. Similarly, Hudson et al. (2011) reported that 13% of their study's participants showed an initial interest in visiting destinations in Latin America after seeing them in the film *Motorcycle Diaries (*2004 2004*)*. Visual media as such have been widely used by tourism bodies and destination marketers to establish, change, or reinforce the image of a tourist destination (Avraham, 2015; Govers et al., 2007; Pan, 2011; Pan et al., 2016).

Past research, also on religious tourism, has further revealed that religious tourists appear to search for signs and symbolic structures – widespread in religious institutions through narrations and visual media – that create meaning and shape their experiences (Badone, 2007; Selwyn, 1996). Using visual media including TV news and documentaries religious authorities often even develop what Baudrillard (1981, p. 85) calls "strategies of desire", through which tourists are mobilized to visit particular places. The visual projection of material objects such as the fame of miraculous icons inscribe places with fantasies and memories, and motivate people to visit sacred places to encounter these objects (Cloke & Pawson, 2008; Coats, 2011; Selwyn, 1996; Tilley, 2006). Therefore, through the selection of an emblematic trait or historical event (i.e. Virgin Mary apparition) that is provided for consumption (see Cloke & Pawson, 2008; Knudsen, 2011) religious sites have gradually become well-known tourist destinations. Additionally, new religious tourist destinations have been established within the last decades, like the Medjugorje in Serbia (Vukonic, 2006), as a result of the visual projection on TV news of stories of miracles that have happened there.

Visual media: on-site experience

Apart from promoting tourism destinations and increasing the number of visitors at locations associated with films or presented on TV, visual media often also influence tourists' on-site experience and behavior (Iwashita, 2008; Kim, 2010, 2012; Mercille, 2005; Rittichainuwat & Rattanaphinanchai, 2015; Siripis et al., 2013) as they further shape (through destination image) expectations before the actual visit takes place (Font, 1997; Phelps, 1986). For example, Siripis et al. (2013) studied the impact of the film *The Beach (2000)* on Thailand's image and concluded that images or videos displayed in popular media pre-programmed visitors' experience in Thailand. Similarly, Kim (2010) explored the impact of a Korean TV drama series on tourists' on-site experience and found that this TV program inspired film tourists to enhance their tourist experiences by re-enacting scenes from the series. In line with Urry (1990, p. 13) people seek to "experience in reality the pleasurable dramas they have already experienced in their imagination".

This is potentially due to the fact that people give meaning and form to their experiences through visual images, signs, and symbols (Bandura, 2001; Bouiding, 1956; Reijnders, 2011), and popular media often circulate and sustain such images (Browne, Firestone, & Mickiewics, 1994; Pan et al., 2014). Popular media, through visual representation as such, enhance tourists' construction of the gaze in that tourists are searching for already known objects, spots, or settings to feast their eyes on (Urry, 1990). *Lord of the Rings (2003)* film tourists, for example, were searching for settings like hobbit houses and attached intrinsically deeper meanings to the locations depicted in the film, which were experienced "as sacred places rather than as sites of spectacle" (Roesch, 2009, p. 134).

Equally within the religious context, visual media along with myths and narratives seem to create spaces for collective gazing and to pre-program (Urry, 1990) religious tourists' expectations and experiences; "familiar narratives of lost cities, ancient civilizations and spirituality facilitate people's partial understandings of that which may be encountered" (Scarles, 2009, p. 470). Nevertheless, religious sites are not only to be gazed at but are also places of conduct. Moving thus beyond a masculine paradigm of the gaze (Ateljevic, Morgan, & Pritchard, 2007; Pritchard & Morgan, 2000), the destination image building process in the case of religious tourism is (further) supported by visual media which underline the importance of other corporeal experiences and senses such as doing and touching, in generating particular experiences that constitute the core of religion (James, 1902/1960). The visual projection of crawling pilgrims on TV has, for example, such an effect on potential visitors' destination choice and on-site experience (Dubisch, 1995). Equally, participation in religious rituals and religious trips, bestow members with feelings of "communitas" (Turner & Turner, 1978), which denotes an egalitarian association between people who share a liminal state. Such states of togetherness and brotherhood are communicated through visual media that broadcast collective religious performances and interviews of believers sharing their on-site experiences, thus influencing other religious tourists who attempt to seek out similar experiences.

The above review of the literature revealed that visual media play a critical role in both the tourism and the religious context. However, only a limited number of studies have adopted an interdisciplinary (combining both religion and tourism) approach and explored the role of visual media in the religious tourism context. Similar to Siripis et al. (2013) and Kim (2012) visual media are expected in this study to have an impact not only on religious tourists' destination image and intention to visit a destination but also affect their on-site experience (tourists' gaze and practices). Specifically, and based on the above, the study seeks to address the following research objectives: (1) to identify and understand the way visuals influence religious tourists' destination image and site selection; and (2) to identify and explain the way visuals affect religious tourists' on-site experiences. This understanding is critical for religious and church authorities, destination managers, and tourism marketers in order to maximize patronage of the destination and to try to shape religious tourists' destination image and choice through marketing initiatives (Sun, Ryan, & Pan, 2015). It also helps destination marketing organizations (DMOs) to better manage visitors' experience by identifying and taking advantage of powerful visual elements in tourism promotion (Pan et al., 2014).

Methodology

Context and study setting

The research objectives are addressed by drawing on participant observation material and in-depth interviews conducted with tourists on an organized religious tour to the sacred island of Tinos, Greece. Despite its small size (8600 inhabitants) Tinos attracts thousands of visitors annually who flock to see the icon of the Virgin Mary in the Church of Annunciation. The Church of Annunciation (built on the location where the icon was found in 1822) has gained fame after numerous reports of therapeutic miracles fulfilled by the icon of the Virgin Mary of Tinos (Panagia Evangelistria), which nowadays is the most visited religious site in Greece (Dubisch, 1995; Terzidou, Stylidis, & Szivas, 2008). Accordingly, while the vast majority of studies on Christian religious tourism have mainly focused on the views and religious practices of Protestants and Catholics, this study gives attention to the Christian Eastern Orthodox church (see Belhassen, Caton, & Steward, 2008; Coleman, 2004), which ranks second in popularity within the Christian world after the Roman Catholic Church.

The Greek TV market

There are eight private and three public TV stations that broadcast nationwide in Greece. According to data obtained from Nielsen Media Research Greece (2015), people in Greece spent on average 4 hours and 26 minutes watching TV in 2015, while those aged over 65 years old watched on average 7 hours per day. Until recently the most popular TV station was Mega

Channel, with its News program being watched by 20% of the TV viewers (763,000 people). Second was Ant1 News with 18%, and third Alpha TV with 16%. In the years 2014–2015, about 20% of the TV News audience was over 65 years old, 19% were male and between 45–64 years old, and 8% were female and aged between 45–64 years. The Greek State owns three TV stations, with the most popular one having on average 10% of the market share and the least popular one about 2.7% of the total audience share. However, it is notable that on August 15, 2015, which is the national celebration of Virgin Mary, the three State TV stations that broadcast the procession in Tinos achieved a cumulative 20% share of the market. Those who watch the three State TV stations tend to be older and with a lower educational level (Nielsen Media Research Greece, 2015).

Sampling and research methods

Considering that tourists' destination image and on-site experience are highly complex, subjective, and unique to each individual (Kim, 2012), qualitative research was considered the most appropriate to study the phenomena in question. Ethnographic research, in particular, was conducted in two organized religious coach trips (each trip was four days long) to the sacred island of Tinos. Purposive sampling was adopted as research participants were selected from the trip's group members. Information leaflets were distributed to prospective participants at the beginning of the trip explaining the purpose of this research. Confidentiality was assured at all stages and participants were allowed to decide whether they would partake in the study or not. In the first stage of the research, participant observation and informal interviews were conducted, with the role of the the researcher as "participant as observer" being adopted. To avoid issues of reactivity and annoyance, and to gain trust in a natural manner, one of the researchers participated in religious practices alongside group members (Bernard, 2006). The similarity between participants' and the leading researcher's national and religious background facilitated engagement, acceptance, and mutual understanding (Andriotis, 2009). Following ethical approval, only participants who gave their consent were observed. In total, 31 observations were conducted during the two trips and detailed field notes were kept throughout.

In the second stage, 38 semi-structured in-depth interviews were conducted after the end of the trips until data saturation was reached (Guba & Lincoln, 1989). The interviews lasted on average 75 minutes and were conducted after assuring participants' confidentiality and receiving their written consent (Kvale & Brinkmann, 2009). In order to determine the multiple realities, the interview questions were as open-ended and flexible as possible so as to encourage respondents to freely communicate their beliefs, values, and so on, without feeling constrained (Fielding & Thomas, 2008). To answer the study's research questions, two broad interview themes were elaborated, namely: (1) How do visual media, and in particular TV news and documentaries, influence your image of Tinos and its selection?; and (2) How do visual media (TV news and documentaries) influence your on-site experience? Topics of discussion for the first main theme included the time/frequency people spent watching TV documentaries and TV news; the TV programs with religious content that respondents know and systematically watch; selection of religious sites, motivation to visit, and expectations, which were addressed, for example, with the use of opinion questions such as "What is your opinion about the documentaries devoted to the religious places in Greece?", and knowledge questions including "What do you know about the miraculous icon of Tinos?", or behavior questions like "Have you watched any TV documentaries related to the Virgin Mary of Tinos?". For the second interview theme, topics of discussion included, among others, participants' on-site activities, religious objects used, symbolism, which were captured with questions such as "If I followed you in the church, what would I see you doing?"; questions related to feelings including "What were your feelings when you first arrived at the island? At the church?", and sensory questions such as "Did you see/experience familiar objects/things known from TV when you visited Tinos?". Follow-up questions including "When did that happen?" were used particularly to establish rapport, convince respondents to expand their answers, and clarify arising issues (O'Reilly, 2005).

All interviews were digitally recorded, transcribed verbatim, and back-translated into English (Brislin, 1976) by both researchers to ensure consistency in meaning. Pseudonyms were used during the data analysis to ensure anonymity. Religious tourists' age ranged from early thirties to early eighties, with the vast majority of respondents (80%) being women. This is not uncommon as in the Greek culture women usually connect the family with religion through religious institutional performances (Dubisch, 1995). Furthermore, the vast majority of them were traveling in small groups of two to three people, and only one out of five traveled alone. About one third of the study's participants were employed full-time (36%), one third had never worked in the past, and the remaining 30% were pensioners. The vast majority reported having one or two children, and only three people reported having none. Lastly, almost half of them had finished basic education (up to high school), 30% had a higher school degree, and another 20% had a university degree or higher. Thematic analysis was applied to reveal patterns and themes (Boyatzis, 1998). In a similar vein, the field notes were analyzed and coded by both researchers to decrease potential bias. The patterns identified in the field notes were linked to the themes that emerged from the interview transcriptions. Data triangulation is known to minimize individual and methodological biases and increases generalizability of the findings (Decrop, 1999).

Findings

The role of TV news and documentaries in shaping destination image and choice

The first part of the findings discusses how TV news and documentaries shape religious tourists' destination image of Tinos and positively influence their decision to visit the island. Religious and political authorities, in particular, appear to use visual media to shape people's images of Tinos through processes of meaning making, and to enhance national pride and collectiveness (Bandyopadhya, Morais, & Chick, 2008). For example, the most important religious festival in Greece is the "Dormition of the Virgin Mary" that takes place annually on August 15 in Tinos. On that day Greeks and the image of festive Tinos circulate all over Greece with TV news and documentaries focusing on the (a) holy icon, (b) leading politicians, and (c) the crowd of people participating in the celebration. This repeated projection of the same or similar shots of items, people, and scenery has a powerful effect on TV viewers, who accumulate certain images of Tinos (Iwashita, 2008).

The projection of the icon

The state-owned Greek TV channels, in particular, offer live broadcasts of the official procession, which involves carrying the icon of the Virgin Mary (image) from the church to the harbour (Figure 1) while the celebration in Tinos also features in the leading news of all private TV stations in the country (Haland, 2012).

> Aspa, for example argued:I wanted to visit Tinos because the icon was very frequently on TV.

Similar to past research, the visual projection in Greek TV news of the icon of the Virgin Mary of Tinos along with stories of miracles seem to be an important influential factor for people visiting Tinos (Kim et al., 2008;

Figure 1. The official procession, August 15 (authors' image).

Kim & Richardson, 2003). As an interviewee, for example, stated:

> I've heard so much (on TV news) ... that influenced me to come ... of disabled people that ... went and got cured ... and of other problems ... personal problems..." (Nicki).

These images are further reinforced by personal experiences and stories included in documentaries about the Virgin Mary of Tinos. A participant, for example noted that:

> If you watch TV documentaries, then you will see the icon ... and listen to stories about some people who experienced ... miracles ... things like that.

During the interviews most people explained how influential documentaries have been in their choice frequently quoting the following:

> A documentary has increased my curiosity to visit *Tinos*.

Some religious tourists visit Tinos in the "trail of miracles" as they desire to feast their eyes on particular miraculous events or even want to experience a miracle by themselves (i.e. to be cured) (Slater, 1986).

The projection of state and church officials

The destination image of Tinos goes beyond being a miraculous place and Tinos is also considered an important centre of "Greek-ness" as it is believed that the icon helped in the fight for Greek independence in 1821. Helena, for example, proudly commented that:

> Even Kolokotronis [an important Greek General in the fight for Greek independence in 1821] was convinced that since the icon of the Virgin Mary was found, the battle would be victorious.

Accordingly, another focal point of the images projected on TV news is the government cabinet members and the head of the Greek Orthodox Church that depict the close relationship between religion and the state. Most TV channels broadcast politicians' on-site speeches in which they annually highlight the contribution of the Church to Greece's independence and pay respect to this institution.

Figure 2. Church of Annunciation – navy officials (authors' images).

For example, Prokopis Paulopoulos, the President of Greece, in his recent public speech in Tinos on August 15 2015 broadcast by most Greek TV stations, attributed the peace in the country to the Virgin Mary's grace. This image-making process is further reinforced by scenes including Greek flags fluttering all over the island of Tinos along with the presence of navy generals and military ships that occasionally fire a gun in a symbolic manner (Figure 2). Visual media as such are often used by authorities including politicians, church leaders, and place marketers to shape people's imagery of places and to direct their gaze (Scarles, 2009) as well as to reinforce national identity and encourage visiting (Beeton, 2005; Hudson et al., 2011). Through its visual representation in popular media as the "Centre of Greece" Tinos has developed a destination image that is linked to "Greek-ness" and religious identity:

> ... here is our Virgin Mary, she is our Virgin Mary ... for those of us who cannot travel outside Greece, the Virgin Mary is Tinos. (Anna)

A visit to Tinos as such is highly related to nationalism (Pretes, 2003), part of being Greek and Christian Orthodox; as Vaso said:

> it is a priority and commitment

for all Greeks. Maria equally stated that:

> every Christian comes round to Tinos. Whoever you may ask, he will say that at least once in his lifetime he has been there.

The Tinos phenomenon, therefore, resembles what Bhardwaj (1973) termed "national pilgrimage", uniting Greek Orthodox people from around the world.

The projection of the crowds

Tinos also acquires the image of a vibrant place with ongoing religious action and collective performances that lead to feelings of togetherness, what Turner and Turner (1978) term "communitas" as a result of sharing mutual feelings, meanings, and experiences within a liminal state. The visual projection on TV of crowds of people queuing, waiting patiently to venerate the icon, seem to increase people's awareness, further cultivate their curiosity and desire to verify the very picture and to be part of it, and highlight Tinos' significance as a religious centre.

> I was moved ... it was so crowded, and all these people are really eager to visit Tinos ... I feel proud to be hristian Orthodox and to be Greek. (Babis).

Religious tourists' performances thus seem to become important elements of authorities' "strategies of desire" as they constitute lived expressions of holiness (Baudrillard, 1981, p. 85). Especially the crowd's religious on-site practices such as crawling, which are projected by both public and private national TV channels, shape Tinos' image and reputation (Figure 3). Makis, for example, said:

> First and foremost it is curiosity that attracts you, because you watch it on television ... you see how they crawl ... on their knees ... and you want to be there on that day when the masses ascend to the church to see what exactly is happening ...

Similarly, Maria stated that:

> My decision to come was mainly out of curiosity ... to experience what I was watching on TV ... people crawling.

People seem to choose to visit Tinos to evidence what they have previously watched on TV, a finding similar to Hudson et al.'s (2011) study among others. This curiosity is further reinforced by the projection on TV of Greek celebrities who attend the August 15 celebration in Tinos. Anna for example argued that:

> I was impressed when I saw a famous singer on TV crawling to the church ... You see, all that influences you.

Figure 3. People crawling – queuing (authors' images).

This result corroborates Yen and Teng (2015), who also reported that celebrity involvement in visual media is positively related to people's intention to visit a destination.

Visual media in the context of religious tourism as such create or reinforce people's awareness and increase their intention to visit sacred places. These results are in line with tourism research which found that people who frequently watch TV programs showing positive aspects of a destination had stronger intentions to visit the projected place (Beeton, 2005; Kim et al., 2007; Kim & Assaker, 2014; Pan & Tsang, 2014). However, the use of visual media by Greek politicians, Church officials and place marketers was acknowledged by some participants like Barbara who stated the following:

> It is the advertising ... everything what we hear on August 15 August, of people going there and worshipping ...I have the feeling that everything starts there, from the media ... Because otherwise who would know about the Virgin Mary of Tinos?

Some people thus seem to be aware of a planned attempt to shape their destination image of Tinos, a finding that appears to partially challenge Gartner's (1993) classification of visual media like TV news and documentaries as autonomous image formation agents, which are perceived as unbiased by consumers.

The role of TV news and documentaries in shaping religious tourists' on-site experience

On the trail of miracles

Apart from shaping people's destination image of Tinos, increasing their awareness, and creating a strong desire to visit the island, TV news and documentaries were found in this study to frame visitors' understanding of the place's sacredness (Badone, 2007; Roesch, 2009) and to influence their expectations and on-site experience (Kim, 2012; Siripis et al., 2013). Religious tourists in Tinos were observed, for example, searching for signs where the Virgin Mary's miraculous power can be manifested (Urry, 1990). While entering the church most religious tourists were seeking for particular objects that were hanging from the ceiling or were positioned in specific special posts in the church. One such notable object, well-projected on TV documentaries and often recalled by religious tourists during the interviews, is the metallic orange tree in the church's entrance. As Tasia stated:

> the orange tree is so beautiful ... it stands on the right side ... a gift to the Virgin Mary from a blind man who was cured. He promised to bring to the Church the first thing he would be able to see, which was the orange tree in his garden.

Popular TV programs and documentaries about the Virgin Mary of Tinos seem to direct peoples' gazes and meanings, with religious tourists engaging in practices of "hunting" and "capturing" religious signs, similar to MacCannell's (1976) and Urry's (1990) semiotic tourists, who are searching for and are influenced by imposed signs in material landscapes.

Additionally, other objects that at first glance seemed to be of secondary value were found to have a considerable impact on people's on-site experience. The suspended oil-lamps and the ornaments in the church and around the icon are part of people's offerings, the so called *tamata*, to the Virgin Mary of Tinos (Figure 4). Each offering is known to depict a miracle experienced by a person and as such the numerous offerings in the church represent the many people who have witnessed miracles in the past. The constant projection in TV documentaries of these religious tourists' offerings seems to be part of an effort to further shape or reinforce the image of Tinos as a sacred place and to influence on-site experiences (Edensor, 2001; Franklin, 2004). Firstly, the projection of offerings on

Figure 4. The icon covered with offerings (authors' image).

TV allows people to reflect on their imperfect nature. As Katia stated:

> ... looking at the many offerings, I thought: oh my God, how many people are suffering ...

Secondly, offerings were found to elevate the significance of Tinos in the eyes of religious tourists; since they are given as a thanksgiving, their presence and projection on TV serves as proof of miracles accomplished by the Virgin Mary of Tinos and highlights Tinos' image as a place of salvation (Selwyn, 1996; Tilley, 2006). As Danay said:

> I saw various things ... Hanging ships, which are said to be offerings made by fishermen ... And I saw also gold jewellery ... and I thought that the people who brought all those things to the church ... their wish must have come true.

In Weber's (1946) words it seems to be an "emotional propaganda". The constant projection of the offerings in documentaries promotes the Virgin Mary's miraculous power, thus enticing people to look at them and to engage in on-site practices like that of wish-making (vow). Katerina, for example, seeing the icon and the offerings also made a vow. As she recounted:

> In fact, when I went there [in the church] I was so much ... I mean I was enticed by the various ... the crowd I saw ... the offerings that were all around the icon, the golden ones, and all those things ... that I made a vow too ...

This process resembles Bourdieu's (1991) institutionally structured performance of gift exchange that intends to safeguard religious institutions' existence. However, the role of offerings in the production of religious tourists' experience was criticized by some study's participants who emphasized the provocative effect the displayed offerings has on people, diminishing the real meaning and value of the icon. Janna, for example, recounted how she walked away from the sight of offerings hanging on the miraculous icon, and chose another non-distracting place in the church:

> I walked away from the icon of the Virgin Mary ... because you cannot see her there, so covered with offerings it is. To be honest – God forgive me – but aren't so many offerings really a planned enticement for the pilgrims?

Projected images on TV news of the icon, the Church of Annunciation, and of religious tourists' ascendance to the church seem to be so powerful that they set expectations, cultivate people's feelings of awe, and influence the experience while being there (Kim, 2010).

Positive influences: confirmation of expectations

For some people a visit to Tinos aimed to re-enliven and reconfirm the visual conceptions they have already acquired through TV (Crouch, 2001; Kim, 2012; Scarles, 2009; Siripis et al., 2013). For example, a fraction of respondents mentioned during the interviews that they have experienced on site exactly what they have seen on TV news and documentaries:

> I had imagined a large icon with many offerings ... an icon covered by offerings ... and large crowds of people worshipping ... I found exactly what I have watched on TV ... (Mairy).

Similarly, Elsa stated that:

> The church was exactly as I imagined it to be and as we watched it on TV. With many stairs, crowded, many candles ...

Similar to the work of Riley et al. (1998) and Kim (2010) on film tourism, TV documentaries appear in this study to encourage a bond between religious tourists and sacred places and enhance the image of Tinos. In line with past research, for those participants who have their expectations confirmed, a higher level of satisfaction with the trip is to be anticipated (Carl, Kindon, & Smith, 2007).

Neutral influences: limited capacity of the visuals

For some other participants though, visual media offer narrow viewpoints and are sometimes inadequate in capturing the atmosphere people actually encounter in places as part of their immaterial and material engagements (Buchmann, Moore, & Fisher, 2010). Kostas, for instance, argued that although he had seen the church several times on TV, what he experienced in Tinos was very intense: "

> This has nothing to do with what you have seen. When you enter the church you feel ... completely different from watching it on TV.

This often results in unexpected and/or unpredicted experiences; as Aspa described it:

> ... when you enter the church you shudder ...

while Niky stated:

> You feel awe ... You are standing next to the Virgin Mary.

It appears that spaces of uncertainty and of "not knowing" occur (Scarles, 2009) as visual representations of the Tinos experience on TV are insufficient to embrace the extraordinary qualities the divine possess. As Yannis commented:

> In the beginning because you go there with ... with all those feelings generated in you beforehand ... you go in a different way ... it's like a fear I can even say of what you will encounter ....

Exaggerated influences: mismatch between expectations and reality

The third stream of participants expressed their disappointment with what they actually experienced as contrasted with what they have seen in documentaries, potentially because

> ... on TV they exaggerate ... (Soula).

Visualized representations of the Church of Annunciation and of the icon have resulted in high expectations and have shaped imaginations of Tinos, a notion that Appadurai (1990) termed "imaged worlds" and Shields (1991) "place-myths". One of the participants, for example, stressed the following:

> It was not as ... I thought that I will see something larger in size ... perhaps, I do not know, due to publicity. The interior was like in a common church, nothing special.

This finding contradicts the study of Iwashita (2008) who reported that several respondents expressed a feeling of *déjà vu* (Iwashita, 2008). Apart from the church, the icon was also less impressive in comparison to what was expected by some religious tourists:

> I didn't perceive it to be as ... I was expecting something different [about the icon] ... It seemed to me very simple. From what I have seen on TV programs I was expecting a large icon, with gold, whereas what I found was a small, common icon.

It appears that "imaged worlds", a combination of images and imaginations, not always reflect the actual reality to be experienced (Kim, 2012).

Conclusion

This study explored the role of TV news and documentaries in religious tourists' destination image, choice, and on-site experience, with a specific reference to the Greek Orthodox context. The findings revealed that the repeated visual projection on TV of the icon of the Virgin Mary, the leading politicians, and of the crowds of visitors on TV news has positive effects on religious tourists' image of the island of Tinos, Greece. The image of Tinos as a sacred place seems to be further reinforced by documentaries presenting people's personal accounts of their own supernatural experiences, positively influencing other religious tourists to visit the island (see Kim et al., 2008; Kim & Richardson, 2003). By applying "strategies of desire" such as the miraculous icon (Baudrillard, 1981), church authorities and place marketers seem to create in religious tourists feelings of curiosity which in turn increase their willingness to visit the island. Additionally, political authorities through visual media have reinforced the image of Tinos as the "Centre of Greece" – highly related to nationalism and Greek Christian Orthodoxy (Beeton, 2005; Scarles, 2009). Lastly, visitors themselves were found to strengthen the image of Tinos, with study participants being positively influenced (in their decision to visit Tinos) by the projected crowds of people and their on-site practices like crawling. Similar to leisure tourists, religious tourists thus seem to choose to visit destinations to evidence what they have previously watched on TV (Hudson et al., 2011). Overall, TV news and documentaries appear to shape people's interest in certain destinations by providing them with myths, images, and emotions, all of which are critical motivators for people to travel (Iwashita, 2008). This finding corroborates previous research on the positive effect visual media have on tourists' destination image and choice (Gong & Tung, 2016; Hudson et al., 2011; Pan, 2011; Shani et al., 2010; Wong & Lai, 2013). Although this link is not new, it has rarely been tested in the context of religious tourism.

The first theoretical contribution of the study is that it extends current knowledge about visual media and tourism by highlighting the influence visuals exercise on religious tourists, who are instructed by religious authorities to visit particular sacred places and to engage in religious performances to generate particular experiences and establish religious belief (Bourdieu, 1991).

Besides shaping destination image and choice, TV news and documentaries were also reported in this study to influence religious tourists' on-site experiences and shape imaginations of what people would experience (Kim, 2012; Siripis et al., 2013; Yen & Teng, 2015). Similar to the findings of Kim (2010) and Roesch (2009), the study's participants were observed re-enacting scenes or searching for signs and objects like the orange tree that have been frequently displayed in TV documentaries. Especially in the case of Tinos, the presence of offerings seemed to further intensify visitors' experience, as these were interpreted as signs of the icon's miraculous power and underlined Tinos' image as a place of salvation. Religious tourists' experiences on site as such seem to have been pre-programmed by the visuals displayed in popular media (Siripis et al., 2013). The current study thus extends past research (Kim, 2010; Scarles, 2009; Siripis et al., 2013; Yen & Teng, 2015) as it was found here that for some people these expectations have been met, while for others visual media exaggerated or were unable to capture the sacred atmosphere that people actually experienced on-site. The study also highlights the role visual media play in shaping – apart from gazing – other corporeal experiences and senses such as touching (Ateljevic et al., 2007; Pritchard & Morgan, 2000). Visual representations of destinations, therefore, affect religious tourists' experience at the destination (Edensor, 2001) with TV documentaries creating a myth of a place (Iwashita, 2008).

The findings of the study have a number of managerial implications for religious authorities and place marketers. First, it becomes evident that the use of an emblematic trait and/or historical event can assist place marketers in developing a strong destination brand. In the case of Tinos the miraculous icon of the Virgin Mary has elevated Tinos' significance not only as a sacred destination but also as the "Centre of Greece". The frequent projection on TV of selected images including the icon, politicians, and the crowd further reinforced this message, especially as autonomous image formation agents including TV news and documentaries have been used. The role of people in the promotion of sacred sites is moreover manifested through the social reproduction of experiences and miracles that trigger visits to sacred places. This can be supplemented in the near future by stories of people included in mini-movies that are known to be effective in capturing tourists' attention and in fostering a positive destination image (Chen, 2015). Second, the displayed offerings along with stories and myths related to them help to increase publicity and shape the identity of Tinos. These offerings were found to greatly influence the on-site experience but should be carefully utilized so as not to create expectations that cannot be met, leading to diminished levels of satisfaction.

Despite its contributions, this study does have limitations. First, the study was conducted in a single destination – more research is needed in multiple settings to further validate the study findings. Second, the findings are specific to religious tourists in Greece and cannot be generalized to other religious persuasions. In line with Kim et al. (2008) the consumption of popular media and tourist experiences are contextualized within different social, cultural, and religious values. Third, as only people who have visited the island of Tinos participated in the study, future research should also focus on the role visual media play in non-visitors' images of Tinos. Studies need also to consider the power relations between religious church authorities and destination management organizations as such knowledge will assist in understanding how dominant images projected in TV documentaries and the news are being produced.

Disclosure statement

No potential conflict of interest was reported by the authors.

References

Andriotis, K. (2009). Sacred site experience: A phenomenological study. *Annals of Tourism Research, 36*(1), 64–84. doi:10.1016/j.annals.2008.10.003

Andriotis, K. (2011). Genres of heritage authenticity: Denotations from a pilgrimage landscape. *Annals of Tourism Research, 38*(4), 1613–1633. doi:10.1016/j.annals.2011.03.001

Appadurai, A. (1990). Disjuncture and difference in the global cultural economy. In M. Featherstone (Ed.), *Global culture: Nationalism, globalization and modernity* (pp. 295–310)). London: Sage Publications.

Ateljevic, I., Morgan, N., & Pritchard, A. (2007). *The critical turn in tourism studies: Innovative research methodologies.* London: Elsevier.

Avraham, E. (2015). Destination image repair during crisis: Attracting tourism during the Arab Spring uprisings. *Tourism Management, 47,* 224–232. doi:10.1016/j.tourman.2014.10.003

Badone, E. (2007). Echoes from Kerizinen: Pilgrimage, narrative, and the construction of sacred history at a Marian

shrine in northwest France. *Journal of the Royal Anthropological Institute, 13*, 453–470. doi:10.1111/j.1467-9655.2007.00436.x
Baerenholdt, J., Haldrup, M., Larsen, J., & Urry, J. (2004). *Performing tourist places*. Aldershot: Ashgate.
Bailey, J. M., & Sood, J. (1993). The effects of religious affiliation on consumer behavior: A preliminary investigation. *Journal of Managerial Issues, 5*(3), 328–352.
Baloglu, S. (2001). Image variations of Turkey by familiarity index: Informational and experiential dimensions. *Tourism Management, 22*, 127–133. doi:10.1016/S0261-5177(00)00049-2
Bandura, A. (2001). Social cognitive theory of mass communication. *Media Psychology, 3*(3), 265–299. doi:10.1207/S1532785XMEP0303_03
Bandyopadhya, R., Morais, D. B., & Chick, G. (2008). Religion and identity in India's heritage tourism. *Annals of Tourism Research, 35*(3), 790–808. doi:10.1016/j.annals.2008.06.004
Baudrillard, J. (1981). *For a critique of the political economy of the sign*. St. Luis, MO: Telos Press.
Beeton, S. (2005). *Film-induced tourism*. Clevedon: Channel View Publications.
Belhassen, Y., Caton, K., & Steward, W. (2008). The search for authenticity in the pilgrim experience. *Annals of Tourism Research, 35*(3), 668–689. doi:10.1016/j.annals.2008.03.007
Bell, D. A. (2003). Mythscapes: Memory, mythology, and national identity. *British Journal of Sociology, 54*(1), 63–81. doi:10.1080/0007131032000045905
Bernard, H. R. (2006). *Research methods in anthropology: Qualitative and quantitative approaches* (4th ed.). Oxford: Altamira Press.
Bhardwaj, S. M. (1973). *Hindu places of pilgrimage in India: A study in cultural geography*. Berkeley: University of California Press.
Bolen, J. S. (1994). *Crossing to Avalon: A Woman's midlife pilgrimage*. San Francisco, CA: Harper.
Boorstin, D. (1964). *The image: A guide to Pseudo-Events in America*. New York, NY: Harper and Row.
Bouiding, K. (1956). *The image*. Ann Arbor, MI: University of Michigan.
Bourdieu, P. (1991). Genesis and structure of the religious field. *Comparative Social Research, 13*, 1–44.
Boyatzis, R. E. (1998). *Transforming qualitative information: Thematic analysis and code development*. Thousand Oaks, CA: Sage.
Brislin, R. W. (1976). Comparative research methodology: Cross-cultural studies. *International Journal of Psychology, 11*, 215–229. doi:10.1080/00207597608247359
Browne, D. R., Firestone, C. M., & Mickiewics, E. (1994). *Television/radio news and minorities*. Queenstown, MD: Aspen Institute.
Buchmann, A., Moore, K., & Fisher, D. (2010). Experiencing film tourism: Authenticity and fellowship. *Annals of Tourism Research, 37*(1), 229–248. doi:10.1016/j.annals.2009.09.005
Carl, D., Kindon, S., & Smith, K. (2007). Tourists' experience of film locations: New Zealand as 'Middle-Earth'. *Tourism Geographies, 9*(1), 49–63. doi:10.1080/14616680601092881
Chen, T. (2015). The persuasive effectiveness of mini-films: Narrative transportation and fantasy proneness. *Journal of Consumer Behaviour, 14*, 1. doi:10.1002/cb.1494
Chi, C. G. Q., & Qu, H. (2008). Examining the structural relationships of destination image, tourist satisfaction and destination loyalty: An integrated approach. *Tourism Management, 29*(4), 624–636. doi:10.1016/j.tourman.2007.06.007
Cloke, P., & Pawson, E. (2008). Memorial trees and treescape memories. *Environment and Planning D: Society and Space, 26*, 107–122. doi:10.1068/d79j
Coats, C. (2011). The melodramatic structure of new age tourist desire. *Tourist Studies, 11*(3), 197–213. doi:10.1177/1468797611432037
Cohen, A. B., & Rankin, A. (2004). Religion and the morality of positive mentality. *Basic and Applied Social Psychology, 26*(1), 45–57. doi:10.1207/s15324834basp2601_5
Cohen, E. (1998). Tourism and religion: A comparative perspective. *Pacific Tourism Review, 2*, 1–10.
Cohen-Ioannides, M. W., & Ioannides, D. (2006). Global Jewish tourism: Pilgrimages and remembrance. In D. Olsen & D. J. Timothy (Eds.), *Tourism religion and spiritual journeys* (pp. 156–171). London: Routledge.
Coleman, S. (2004). From England's Nazareth to Sweden's Jerusalem: Movement, (virtual) landscapes and pilgrimage. In S. Coleman & J. Eade (Eds.), *Reframing pilgrimage: Cultures in motion* (pp. 45–68). London: Routledge.
Crang, P. (1997). Performing the tourist product. In C. Rojek & J. Urry (Eds.), *Touring cultures: Transformations of travel and theory* (pp. 137–154). London: Routledge.
Crouch, D. (2001). Spatialities and the feeling of doing. *Social & Cultural Geography, 2*(1), 61–75. doi:10.1080/14649360020028276
Decrop, A. (1999). Triangulation in qualitative tourism research. *Tourism Management, 20*, 157–161. doi:10.1016/S0261-5177(98)00102-2
Diaries, Motorcycle. (2004). *Directed by walter salles*. UK: FilmFour.
Dubisch, J. (1995). *In a different place: Pilgrimage, gender, and politics at a Greek Island Shrine*. Princeton, NJ: Princeton University Press.
Edensor, T. (2001). Performing tourism, staging tourism: (Re) Producing tourist space and practice. *Tourist Studies, 1*(1), 59–81. doi:10.1177/146879760100100104
Eidelson, R., & Eidelson, J. I. (2003). Dangerous ideas: Five beliefs that propel groups toward conflict. *American Psychologist, 58*(3), 182–192. doi:10.1037/0003-066X.58.3.182
Eliade, M. (1969). *Images and symbols*. New York, NY: Sheed and Ward.
Fakeye, P. C., & Crompton, J. L. (1991). Image differences between prospective, first-time and repeat visitors to the lower Rio Grande Valley. *Journal of Travel Research, 30*(2), 10–16. doi:10.1177/004728759103000202
Fielding, N., & Thomas, H. (2008). Qualitative interviewing. In N. Gilbert (Ed.), *Researching social life* (3rd ed., pp. 245–265). London: Sage.
Fleischer, A., & Pizam, A. (2002). Tourism constraints among Israeli seniors. *Annals of Tourism Research, 29*(1), 106–123. doi:10.1016/S0160-7383(01)00026-3
Font, X. (1997). Managing the tourist destination's image. *Journal of Vacation Marketing, 3*(2), 123–131. doi:10.1177/135676679700300203
Franklin, A. (2004). Tourism as an ordering: Towards a new ontology of tourism. *Tourist Studies, 4*(3), 277–301. doi:10.1177/1468797604057328

Gartner, W. C. (1993). Image formation process. *Journal of Travel and Tourism Marketing, 2*(2/3), 191–215. doi:10.1300/J073v02n02_12

Geering, L. (2002). *Christianity without god*. Wellington: Bridget Williams Books.

Geertz, C. (1973). *The interpretation of cultures*. New York, NY: Basic Books.

Gong, T., & Tung, V. W. S. (2016). The impact of tourism minimovies on destination image: The influence of travel motivation and advertising disclosure. *Journal of Travel & Tourism Marketing*. doi:10.1080/10548408.2016.1182458

Govers, G., Go, F., & Kumar, K. (2007). Promoting tourism destination image. *Journal of Travel Research, 46*, 15–23. doi:10.1177/0047287507302374

Graburn, N. H. (1989). Tourism: The sacred journey. In V. Smith (Ed.), *Hosts and guests* (pp. 21–36). Philadelphia: University of Pennsylvania Press.

Guba, E. G., & Lincoln, Y. S. (1989). *Fourth generation evaluation*. London: Sage.

Haland, E. J. (2012). The Dormition of the Virgin Mary on the Island of Tinos: A performance of gendered values in Greece. *Journal of Religious History, 36*(1), 89–117. doi:10.1111/j.1467-9809.2011.01157.x

Hudson, S., Wang, Y., & Gil, S. M. (2011). The influence of a film on destination image and the desire to travel: A cross cultural comparison. *International Journal of Tourism Research, 13*(2), 177–190.

Iwashita, C. (2008). Roles of films and television dramas in international tourism: The case of Japanese tourists to the UK. *Journal of Travel & Tourism Marketing, 24*(2/3), 139–151. doi:10.1080/10548400802092635

James, W. (1902). *The varieties of religious experience* (pp. 1960). New York, NY: Collier.

Kaelber, L. (2006). Place and pilgrimage: Real and imagined. In W. H. Swatos (Ed.), *On the road to being there: Studies in pilgrimage and tourism* (pp. 277–295). Leiden: Brill.

Kim, H., & Richardson, S. L. (2003). Motion picture impacts on destination images. *Annals of Tourism Research, 30*(1), 216–237. doi:10.1016/S0160-7383(02)00062-2

Kim, S. (2010). Extraordinary experience: Re-enacting and photographing at screen-tourism locations. *Tourism and Hospitality Planning & Development, 7*(1), 59–75. doi:10.1080/14790530903522630

Kim, S. (2012). Audience involvement and film tourism experiences: Emotional places, emotional experiences. *Tourism Management, 33*, 387–396.

Kim, S., Agrusa, J., Chon, K., & Cho, Y. (2008). The effects of Korean pop culture on Hong Kong residents' perceptions of Korea as a potential tourist destination. *Journal of Travel & Tourism Marketing, 24*(2–3), 163–183. doi:10.1080/10548400802092684

Kim, S., Agrusa, J., Lee, H., & Chon, K. (2007). Effects of Korean television dramas on the flow of Japanese tourists. *Tourism Management, 28*(5), 1340–1353. doi:10.1016/j.tourman.2007.01.005

Kim, S., & Assaker, G. (2014). An empirical examination of the antecedents of film tourism experience: A structural model approach. *Journal of Travel & Tourism Marketing, 31*(2), 251–268. doi:10.1080/10548408.2014.873316

Kim, S., Kim, M., Agrusa, J., & Lee, A. (2012). Does a food themed TV drama affect perceptions of national image and intention to visit a country? An empirical study of Korea TV drama. *Journal of Travel & Tourism Marketing, 29*(4), 313–326. doi:10.1080/10548408.2012.674869

Knudsen, B. T. (2011). Thanatourism: Witnessing difficult pasts. *Tourist Studies, 11*(1), 55–72. doi:10.1177/1468797611412064

Kotler, P., Haider, D. H., & Rein, I. (1993). *Marketing places: Attracting investment, industry, and tourism to cities, states, and nations*. New York, NY: The Free Press.

Kvale, S., & Brinkmann, S. (2009). *Interviews: Learning the craft of qualitative research interviewing*. London: Sage.

Lanczkowski, G. (1982). *Die heilige Reise: Auf den Wegen von Gottern und Menschen*. Freiburg: Herder.

Lee, C. K., Lee, Y. K., & Lee, B. (2005). Korea's destination image formed by the 2002 world cup. *Annals of Tourism Research, 32*(4), 839–858. doi:10.1016/j.annals.2004.11.006

Lord of the Rings (2003). *Directed by peter jackson*. United States: WingNut Films

MacCannell, D. (1976). *The tourist: A new theory of the leisure class*. New York, NY: Schocken.

Mattila, A., Apostolopoulos, Y., Somnez, S., Yu, L., & Sasidharan, V. (2001). The impact of gender and religion on college students' spring break behaviour. *Journal of Travel Research, 40*(2), 193–200. doi:10.1177/004728750104000210

Mercille, J. (2005). Media effects on image: The case of Tibet. *Annals of Tourism Research, 32*(4), 1039–1055. doi:10.1016/j.annals.2005.02.001

Morgan, N., Pritchard, A., & Piggott, R. (2002). New Zealand, 100% pure. The creation of a powerful niche destination brand. *Journal of Brand Management, 9*(4), 335–354. doi:10.1057/palgrave.bm.2540082

Nielsen Media Research Greece 2015. *Annual report 2015*. AGB Nielsen: Athens.

O'Reilly, K. (2005). *Ethnographic methods*. London: Routledge. doi:

Pan, S. (2011). The role of TV commercial visuals in forming memorable and impressive destination images. *Journal of Travel Research, 50*(2), 171–185. doi:10.1177/0047287509355325

Pan, S., Lee, J., & Tsai, H. (2014). Travel photos: Motivations, image dimensions, and affective qualities of places. *Tourism Management, 40*, 59–69. doi:10.1016/j.tourman.2013.05.007

Pan, S., Santos, C., & Kim, S. (2016). Promoting tourism, projecting power: The role of television commercials. *Journal of Travel & Tourism Marketing*. doi:10.1080/10548408.2016.1156610

Pan, S., & Tsang, N. (2014). Inducible or not—A telltale from two movies. *Journal of Travel & Tourism Marketing, 31*(3), 397–416. doi:10.1080/10548408.2014.883345

Phelps, A. (1986). Holiday Destination Image- the problems of assessment: An example developed in Menorca. *Tourism Management, 7*(3), 168–180. doi:10.1016/0261-5177(86)90003-8

Poria, Y., Butler, R., & Airey, D. (2003). Tourism, religion and religiosity: A holy mess. *Current Issues in Tourism, 6*(4), 340–363. doi:10.1080/13683500308667960

Poulson, R. L., Eppler, M. A., Satterwhite, T. N., Wuensch, K. L., & Bass, L. A. (1998). Alcohol consumption, strength of religious beliefs, and risky sexual behavior in college students. *Journal of American College Health, 46*(5), 227–232. doi:10.1080/07448489809600227

Pretes, M. (2003). Tourism and nationalism. *Annals of Tourism Research, 30*(1), 125–142. doi:10.1016/S0160-7383(02)00035-X

Pritchard, A., & Morgan, N. (2000). Privileging the male gaze: Gendered tourism landscapes. *Annals of Tourism Research, 27*(4), 884–905. doi:10.1016/S0160-7383(99)00113-9

Qu, H., Kim, L. H., & Im, H. H. (2011). A model of destination branding: Integrating the concepts of the branding and destination image. *Tourism Management, 32*(3), 465–476. doi:10.1016/j.tourman.2010.03.014

Reijnders, S. (2011). *Places of the imagination: Media, tourism, culture*. Surrey, UK: Ashgate Publishing.

Riley, R., Baker, D., & van Doren, C. S. (1998). Movie induced tourism. *Annals of Tourism Research, 25*(4), 919–935. doi:10.1016/S0160-7383(98)00045-0

Rinschede, G. (1992). Forms of religious tourism. *Annals of Tourism Research, 19*, 51–67. doi:10.1016/0160-7383(92)90106-Y

Rittichainuwat, B., & Rattanaphinanchai, S. (2015). Applying a mixed method of quantitative and qualitative design in explaining the travel motivation of film tourists in visiting a film-shooting destination. *Tourism Management, 46*, 136–147. doi:10.1016/j.tourman.2014.06.005

Roesch, S. (2009). *The experiences of film location tourists*. Bristol: Channel View Publications.

Scarles, C. (2009). Becoming tourist: Renegotiating the visual in the tourist experience. *Environment & Planning D: Society and Space, 27*(3), 465–488. doi:10.1068/d1707

Schofield, P. (1996). Cinematographic images of a city. *Tourism Management, 17*(5), 333–340. doi:10.1016/0261-5177(96)00033-7

Selwyn, R. (1996). *The tourist image*. London: Wiley.

Shani, A., Chen, P. J., Wang, Y., & Hua, N. (2010). Testing the impact of a promotional video on destination image changes: Application of China as a tourism destination. *International Journal of Tourism Research, 12*, 116–133.

Shields, R. (1991). *Places on the margin: Alternative geographies of modernity*. London: Routledge.

Silberman, I. (2005). Religion as a meaning system: Implications for the New Millennium. *Journal of Social Issues, 61*(4), 641–663. doi:10.1111/josi.2005.61.issue-4

Singh, S. (2006). Tourism in the sacred Indian Himalayas: An incipient theology of tourism? *Asia Pacific Journal of Tourism Research, 11*(4), 375–389. doi:10.1080/10941660600931226

Siripis, M., Scarles, C., & Airey, D. (2013). Being a tourist or a performer? Tourists' negotiation with mediated destination image in popular film. In J. A. Lester & C. Scarles (Eds.), *Mediating the tourist experience: From brochures to virtual encounters*. London: Routledge.

Slater, C. (1986). *Trail of miracles: Stories from a pilgrimage in Northern Brazil*. Berkeley: University of California Press.

Smith, V. (1992). The quest in guest. *Annals of Tourism Research, 19*, 1–17. doi:10.1016/0160-7383(92)90103-V

Stylidis, D., Belhassen, Y., & Shani, A. (2015). Destination image, on-site experience and behavioural intentions: Path analytic validation of a marketing model on domestic tourists. *Current Issues in Tourism*. doi:10.1080/13683500.2015.1051011

Stylidis, D., Shani, A., & Belhassen, Y. (2017). Testing an integrated destination image model across residents and tourists. *Tourism Management, 58*, 184–195. doi:10.1016/j.tourman.2016.10.014

Sun, M., Ryan, C., & Pan, S. (2015). Using Chinese travel blogs to examine perceived destination image: The case of New Zealand. *Journal of Travel Research, 54*(4), 543–555. doi:10.1177/0047287514522882

Terzidou, M., Stylidis, D., & Szivas, E. (2008). Residents' perceptions of religious tourism and its socio-economic impacts on the island of Tinos. *Tourism and Hospitality: Planning and Development, 5*(2), 113–129.

The Beach (2000). Directed by Danny Boyle. UK: 20th Century Fox.

Tilley, C. (2006). Introduction: Identity, place, landscape and heritage. *Journal of Material Culture, 11*(1/2), 7–32. doi:10.1177/1359183506062990

Timothy, D. J., & Iverson, T. (2006). Tourism and Islam: Considerations of culture and duty. In D. J. Timothy & D. H. Olsen (Eds.), *Tourism, religion and spiritual journeys* (pp. 186–205). London: Routledge.

Turner, V. (1973). The centre out there: Pilgrim's goals. *History of Religion, 12*(3), 191–230. doi:10.1086/462677

Turner, V., & Turner, E. (1978). *Image and pilgrimage in christian culture: Anthropological perspectives*. New York, NY: Columbia University Press.

Urry, J. (1990). *The tourist gaze: Leisure and travel in contemporary societies*. London: Sage.

Vukonic, B. (2006). Sacred places and tourism in Roman Catholic tradition. In D. J. Timothy & D. H. Olsen (Eds.), *Tourism, religion and spiritual journeys* (pp. 237–253)). London: Routledge.

Weber, M. (1946). The social psychology of the world religions. In H. H. Gerth & C. W. Mills (Eds.), *From Max Weber: Essays in sociology* (pp. 267–301). New York, NY: Oxford University.

Williams, E., Francis, L., Robbins, M., & Annis, J. (2007). Visitor experiences of St Davids Cathedral: The two worlds of pilgrims and secular tourists. *Rural Theology, 5*, 111–123. doi:10.1179/rut_2007_5_2_004

Wong, J. Y., & Lai, T. C. (2013). Celebrity attachment and behavioral intentions: The mediating role of place attachment. *International Journal of Tourism Research, 17*(2), 161–170. doi:10.1002/jtr.1974

Yen, C. H., & Croy, W. G. (2016). Film tourism: Celebrity involvement, celebrity worship and destination image. *Current Issues in Tourism, 19*(10), 1027–1044. doi:10.1080/13683500.2013.816270

Yen, C. H., & Teng, H. Y. (2015). Celebrity involvement, perceived value, and behavioral intentions in popular media-induced tourism. *Journal of Hospitality & Tourism Research, 39*(2), 225–244. doi:10.1177/1096348012471382

Reality television portrayals of Kavos, Greece: tourists behaving badly

Nicola Williams-Burnett, Heather Skinner and Julia Fallon

ABSTRACT

Reality television programmes are an increasingly ubiquitous element of popular culture. While other forms of popular culture's impact on tourists' perceptions of destination image have been considered in the extant literature, the impact of reality television is under-researched. This article provides a qualitative thematic analysis of televisual media portrayals of the tourist destination of Kavos, on the Greek island of Corfu, whose economy is almost solely reliant upon an increasingly short high-summer season. Reality television portrayals of tourist behavior have generated negative publicity about the resort's image that may negatively affect potential tourists' perceptions of Kavos as a destination.

Introduction

Tourism as escape from the day to day remains "central to most forms of tourist experience" (Williams & Lew, 2015, p. 12). Wang (1999, p. 363) proposed that tourist motivations can involve self-making and self-identity, particularly for individuals who "cannot realize their authentic selves in everyday life". Young tourists may find such realisation constraints brought about through relatively close parental supervision, and additional opportunities brought about by the amount of free time available to party whilst on holiday compared to when they are at home (Andriotis, 2010). Reports in Greek and British newspapers based on the behaviour of young British tourists on Club 18–30 holidays in Greek island resorts have been overwhelmingly in a "disapproving and moralising tone" (Andriotis, 2010). The popular destination of Kavos, once a relatively unspoilt fishing village at the far southern tip of the Greek island of Corfu, is now perceived as a hedonistic summer party destination particularly for young British tourists, and the Kavos economy is almost solely reliant upon an increasingly short high-summer tourist season when these young people are free particularly from full-time education commitments. Four different series of reality television programmes, that have to a greater or lesser extent included a focus on Kavos, have all concentrated their attention on images of extremes of behaviour of this particular tourist demographic. Since the airing on British television of these programmes the destination image of Kavos is increasingly perceived as negative (Bastakis, Buhalis, & Butler, 2004; Kamenidou, Mamalis, Kokkinis, & Geranis, 2013). Kavos thus finds itself in a group of such destinations that are "in a difficult to reverse position" given the type of tourism they are attracting that have "spoiled heavily their image and reputation" (Bastakis et al., 2004). One study based on questionnaires and interviews in a resort with British tourists holidaying in Kavos in the high-summer season of 2011 has described the resort as a "wild destination", "with intense night life and clubbing", and tourist experiences involving "excessive drinking, drug use, risk sex behaviors and many times the consequences are injuries, accidents, vandalism, fights, assaults, and even deaths" (Kamenidou et al., 2013, pp. 100–102).

The purpose of this paper is to provide an analysis of the destination image of Kavos according to the content of these reality television programmes. This study therefore continues and further contributes to the development of theoretical understanding of the relationship between various forms of popular culture and destination image particularly regarding reality televisual portrayals of popular tourist destinations. Our focus on the way media can present a destination in a way that may lead to a negative image being perceived contributes to an under-developed area in the literature on media-induced tourism. In particular, the contribution of this research has been to further the development of theoretical understanding of the relationship between reality television as a form of popular culture and destination image that has been previously lacking in the literature.

The paper's structure is based on an introduction to the concepts under investigation. This is followed by a review of relevant literature. We then present and defend the methodological approach taken. Findings are then presented, and summarised. These findings are then discussed in light of the literature. Our conclusion highlights the original contribution this paper has made, and also considers the limitations of this study and suggested areas for further research.

In undertaking this research we have employed a highly qualitative methodology, thematically analysing the textual content of each episode of these television programmes. In order to provide a relevant theoretical basis to this paper, first we will review the literature on destination image and media-induced tourism. We then provide insights into the relatively recent, but growing phenomenon that is reality television, because reality television is indeed a different phenomenon to, for example, drama or documentary, and we therefore believe that such a discussion provides a more specific underpinning for this paper that points out the way reality television has so far been largely ignored by researchers of destination image and media-induced tourism. We then present and defend the use of the highly qualitative, and indeed some may claim overly subjective, methodological approach we have used to thematically analyse our data, before presenting our findings. Our findings have been presented firstly programme-by-programme, and this section of our paper concludes with a summary of the key themes arising from our overall analysis. The discussion of our findings highlights the differences between reality television and other forms of media. The findings are thus indeed new and provide a contribution to furthering academic and practitioner understanding of the way different forms of visual media can contribute to destination image formation. Because of the contrived nature of reality television programmes, they do not show the place as a year-round location where people live and work, nor do they portray the broader reality of what really happens in these places even at the height of the season when not all young tourists behave in the extreme ways that make for sensationalised television. Our conclusion therefore highlights our original contribution to knowledge that such reality television portrayals of youth-targeted "party" destinations such as Kavos are highly contrived by those responsible for making these programmes, but are consumed as "reality" by many audiences, and are also consumed differently than travelogues and other documentaries that purport to portray a place's reality, or fictionalised dramas that use a place only as a location backdrop to the programme itself.

Destination image and media-induced tourism

One of the first theories of image formation was proposed by Gunn (1972) who differentiated between *Organic images* (arising from unbiased external sources), *Induced images* (derived from marketing and promotional material emanating from the destination), and *Modified-Induced images* (following first-hand experience of the destination). Fakeye and Crompton (1991) similarly identify three stages of image formation linked to such sources, although they define the final stage as the Complex image. As the terms used by various authors to refer to each of these image formation stages and sources can be confusing depending upon which cited source is used, and as various authors *have unpacked and labelled the various information sources somewhat differently (notably* Gartner, 1993; Echtner & Ritchie, 1991) this article will use the terms relating to the three main sources and stages of image formation consistent with Fakeye and Crompton (1991). This paper is not concerned with *Induced images*, rather, a focus on reality television portrayals of a destination would fall into the category of *Organic image* source materials. There is some reference to the *Complex image* formed from first-hand experience of the destination, as all the television programmes we include in our study show tourists on holiday in-resort.

The tourism literature has long recognised that a destination's image has an effect on the tourist's decision process (Kim & Richardson, 2003), and on tourists' perceptions of places (Lawson & Baud-Bovy, 1977). "Image formation is defined as a construction of a mental representation of a destination on the basis of information cues delivered by the image formation agents and selected by a person" (Tasci & Gartner, 2007, p. 415).

Destination image is defined as "the sum of beliefs, ideas and impressions that a person has of a destination" (Crompton, 1979, p. 18) and destination image serves as a cue and guidance for tourists and can help with the decision-making process when choosing and purchasing a holiday (Morgan & Pritchard, 1998). The tourist industry is a highly competitive market with many different product offerings for the consumer to choose from. Purchasing a holiday often entails a high degree of involvement and risk for the individual as they will invest time, money, and emotion into this process (Blackwell, Miniard, & Engel, 2005). Individuals will utilise a variety of media sources and discussions with friends and family to gain a greater understanding of a destination (Brayshaw, 1995). Therefore, it is crucial that a resort, destination, location, or country creates the right image when promoting itself to prospective tourists (Morgan & Pritchard, 1998). However, destination images are not only *induced* by those charged with the formal task of promoting and encouraging tourism, but are also created through other *organic* sources like friends, family, and media such as film, television, literary works, and online media. Moreover, it is not only tourists and potential tourists who consume such destination images. Particularly with regard to destination images from various forms of visual, textual, and symbolic media representations (Iwashita, 2006), film and television audiences may consume destination images as entertainment, or may come across image formation information in newspapers, and via social media platforms. Even when state-sponsored or destination marketing organisation (DMO)-sponsored messages may be found in these portrayals, (notably, for example, when a nation's Film Board or other similar body provides inducements for a feature film or television programme to be located in the destination) the actual representation of the destination usually remains in the control of the production company.

Karpovich (2010, pp. 9–10) identifies that "the funeral in 1926 of film star Rudolph Valentino, reportedly attended by close to 100,000 fans" can be seen as "one of the earliest recorded examples of the significance of film as an entirely distinct and purposeful motivation for travel". Disneyland, opening in 1955, also provided a tourist experience associated with the film output of the Disney studio, as did the airline El Al's "organised tour of the locations used for the filming of Exodus", and a growth in academic interest in the mid-1990s led to the recognition of "film-motivated tourism as a distinct field of enquiry". Broadening the academic approach to consider other forms of media, Busby and Klug (2001, p. 316) now define "media-induced tourism" as that which "involves visits to places celebrated for associations with books, authors, television programmes and films". "Films, television programmes, and literature have an excellent ability to lead people to a strong interest in a destination by providing them with such imagery, memories, myths, icons, and emotions" (Iwashita, 2006, p. 63), they can influence travel preferences, and indeed, as a promotional method "have become a catalyst for tourism in contemporary globalised media-saturated society" (Iwashita, 2003, p. 332). As a result, a number of studies have analysed the effects of such media-induced tourism, but, in the main, this research has tended to lean more toward its positive effects (Beeton, 2005; Busby & Klug, 2001; Hudson & Ritchie, 2006; Skinner, 2016; Spears, Josiam, Kinley, & Pookulangara, 2013). Although Lichrou, O'Malley, and Patterson (2008, p. 33) recognise that "popular culture can be seen as a medium for the construction of places in the tourists' minds", with television being stated as one such form of popular culture, there is little research related to the effects of a destination's portrayal in reality television programmes at all, and little in the general literature on media-induced tourism that considers any negative effects these programmes may have on the image of a destination.

Reality television

Televisual media content "can reproduce or counter misapprehensions of a place" (McElroy, 2011, p. 176). For a destination such as Kavos where negative and pejorative reports on tourist behaviour have regularly been printed in the press, television programmes focusing on tourists behaving badly in the resort can therefore increase the possibility of the consumers of such media perceiving the destination in a negative light. However, both as a medium and as a technology, television "is undergoing a spatial turn of its own on various levels. Television texts today migrate from one screen to another and from platform to platform … as media content is subject to a 'space shift' … and no longer confined to a 'box' in the living room" (Shagrir, 2015). Much televisual media content is also now freely (if not always legally) available via the Internet, which itself is already increasingly becoming a strong influencing factor on tourists' decisions about destinations (Alonso & Bea, 2012).

One type of televisual media content that is under-researched is that of the reality television programme. There are many different genres of reality television, including documentary-style programmes featuring celebrities (such as *Keeping up with The Kardashians* or *The Osbournes*), special living environments (*Big Brother, Jersey Shore)*, professional activities (*Miami Ink*) and subcultures (*The Undateables*), reality legal programmes (*Judge Judy*), reality competition game shows (*Britain's Got Talent* and *X-Factor*), and self-improvement programmes (*The Biggest Loser*). Nabi, Biely, Morgan, and Stitt (2003, p. 304) define reality television as "programs that film real people as they live out events (contrived or otherwise) in their lives, as these events occur", thus blurring the lines between destination image on television as documentary or as entertainment, although Dunn (2005) even questions the "authenticity" of the way places are portrayed in more traditional documentary/entertainment-style travel/holiday programmes on television. The term "documentary" is attributed to film maker John Grierson, who defined documentary as "the creative treatment of reality" (Winston, 2008). Gilhespy and Harris (2011, p. 53) claim that "documentary film belongs to a discourse that claims to describe the real and to offer insights into the truth", while also warning that editorial inputs into the way such reality is produced creatively tend to be hidden or disguised. Documentary is also at least to some extent scripted, yet differing from other scripted forms of televisual media such as fictional drama inasmuch as documentary retains its claim as a form that "offers accounts of 'the real'" (Gilhespy & Harris, 2011, p. 54).

Nabi et al. (2003) stress that reality television programmes are not scripted, and therein lays the fundamental difference between this study and extant studies focusing on other types of media-induced tourism, particularly scripted films, television dramas, or documentaries. Rose and Wood (2005, p. 284) propose that "the consumption of reality programming represents a sophisticated quest for authenticity within the traditionally fiction-oriented entertainment paradigm", although they do question the ability of such consumption to provide insights into authenticity when programmes are "increasingly characterized by simulation". Research has identified that whilst participants who watched reality programmes understood that the shows are probably in some way staged, participants were likely to view reality television programmes as "more real" than more traditional forms of scripted television programmes and/or films (Nabi et al., 2003).

As reported in the popular press, "the January series of What Happens in Kavos drew in 1.61 m viewers, whilst on BBC Three, the most recent series of Sun, Sex and Suspicious Parents hit 1.1 m viewers. TV voyeurism is compelling" (Volpe, 2013). However, Volpe (2013) goes on to stress that it is the airing of these reality television programmes that has contributed to broadening the audience reach insofar as portraying Kavos unfavourably, based on the "reality" as shown in such programmes:

> Most of us were blissfully ignorant as to what went on ... until luckily (or unfortunately) for us BBC Three and Channel 4 ventured into the bars, clubs and emergency clinics to follow around scores of revellers drinking, having casual sex, and generally making tits of themselves on national television.

Such is the effect of these programmes on the destination image of Kavos in the popular press that one television critic (O'Sullvan, 2013) even ended his critique of the programme thus:

> Footnote: Kavos is on the south coast of Corfu. Don't, whatever you do, go there ... .

Methodology

The focus of our research reflects the growing interest in the visual across a range of disciplines, especially tourism, where the inclusion of such methods is relatively recent due to both "the increasing legitimisation of qualitative research, and the willingness of tourism researchers to explore innovative approaches to research" (Rakić & Chambers, 2012, p. 4).

Much of the extant literature on media-induced tourism has been dominated by quantitative approaches that seek to statistically identify relationships and predictors in relation to its effects (Hudson & Ritchie, 2006; Spears et al., 2013). Moreover, tourism studies are "still predominantly located within business and management schools where visual methods have not been widely legitimated" (Chambers, 2012, p. 33). Thus Skinner (2011) suggests that other qualitative approaches such as those more usually applied within the social sciences should be adopted in order to explore emerging themes within place identity formation, although recognising that such approaches may be less familiar to business and management researchers. Iwashita (2003) favours a social constructionist approach to studying place portrayals in contemporary media. This study leans heavily towards the qualitative approaches of social constructionism in the way tourism consumption practices within Kavos are portrayed in televisual media and the way such practices impact on its destination image.

Andriotis (2010) explored print media attitudes to the behaviours of young people on holiday in Club 18–30 resorts in Greece. Our study has employed a similar analysis, but of the content of the reality television programmes focusing on Kavos. We also take a similar line of defence as did Andriotis (2010) in justifying our focus on this particular destination as an appropriate setting for research due to the amount of media coverage that is given to the behaviour of young British tourists in such Greek resorts.

Within Andriotis' (2010) study, as with ours, the first stage was for the researchers to identify appropriate media for analysis. When considering other forms of visual media in tourism research, issues of subjectivity and objectivity tend to be associated with the visual medium that is being researched. For example, the use of tourist photographs as data, while understood to involve to some extent subjective action on behalf of the photographer, is often associated with positivism because similar to any machine "the photographic camera has been understood as a scientific tool for registering reality more accurately" (Gilhespy & Harris, 2011, p. 52). However, there is also an interesting methodological paradox regarding the way photographs are perceived:

> Viewed as works of art, photographs are thought to embody the personal concerns of the photographer-artist ... viewed as records, photographs are thought to reproduce the reality in front of the camera's lens, yielding an unmediated and unbiased visual report (Schwartz, 1989, p. 120).

Similarly, fictionalised drama could be considered as more of a subjective portrayal of a location than a documentary about a particular place (regardless of editorial input). Reality television, on the other hand, while claiming portrayal of the real in its epithet, is certainly recognised as being a more subjective form of visual media representation than documentary, scripted as it is in a manner more akin to fictionalised drama.

As the purpose of this study is to provide an analysis of the destination image of Kavos according to the content of "reality" television programmes, we set the selection criteria that the reality programmes considered for analysis must be set in an identifiable holiday location, filmed in a documentary style giving the viewer "fly on the wall" viewing, shown on United Kingdom (UK) television, and where none of the cast was classed as a celebrity. Thus data were gathered from these four reality programmes:

- *Greece Uncovered (GU)* – made by London Weekend Television (LWT), with eight episodes filmed in "party" locations around Greece, including Kavos, aired on Sky One between 9 August and 27 September 1998. It was the third of seven of the *Uncovered* series of programmes that in previous years focused on Ibiza and the Caribbean. Further seasons focused on Miami, Tenerife, Ibiza again, and Australia.
- *Sun, Sex and Suspicious Parents (SS&SP)* – featured young holidaymakers on their first vacation away from their parents, with the "twist" that the parents too were in the holiday destination, secretly watching their behaviour. Aired on BBC3 between 2011 and 2014, Kavos was featured in Series 1 (S1) Episode 4 (Ep4) (25 January 2011); S2 Ep3 (8 February 2012); and S3 Ep 6 (12 February 2013).
- *Corfu: A Tale of Two Islands (CTOTI)* – an eight episode programme aired on ITV in October and November 2012 that explored the life of expatriates (expats) and holidaymakers on the island.
- *What Happens in Kavos (WHIK)* – made by Dragonfly for Channel 4. Season 1 aired in January 2013, season 2 in November and December 2013.

Similar to Andriotis (2010) we also found that a quantitative approach to analysing the content of these media portrayals would be "inadequate, mainly because the emphasis of this study was not to obtain data that can be counted and analysed statistically, but to fulfil the goals of the quest for meaning". Thus we explored a more qualitative thematic method of analysis to understand the way reality television programmes represent the destination image of Kavos, with the aim of identifying messages, patterns, and/or themes (Braun & Clarke, 2006; Roberts & Pettigrew, 2007) contained within these programmes, seeking phrases to provide and enhance understanding and perception of the issues (Hanson, 1992).

When considering methodological approaches to qualitative analysis, it must be recognised that such an analysis is subjective. In the opinion of these authors that is simply a characteristic of qualitative research and alone does not, and should not, belie a negative judgement about its worth. Our analysis was undertaken qualitatively, thematically, and not using any form of mechanised processes or software to assist in this analysis. Indeed, Davidson and Skinner (2010) found that there was no benefit to be gained from the use of computer-aided analysis over manual analysis of various forms of qualitative data, further stressing that computer-aided techniques were found to be restricted and highly limited in their utility for analysing various forms of visual data.

We therefore treated the reality television programmes comprising our data set as subjective qualitative data, and each episode of each programme was viewed in order to ascertain the themes arising from these data. Some episodes were viewed more than once in order to check the initial analysis that had been made of its content. Some episodes were viewed again after the entire data set had been viewed and analysed to add more relevant insight into findings once the themes had arisen from the initial viewing. The number of viewings was not consistent across the entire data set because of the subjective nature of this form of qualitative enquiry. Rather each episode was viewed the number of times necessary in order to ascertain the themes arising from the programmes overall.

Our data set comprised each episode of the abovementioned reality television programmes, and our thematic analysis was conducted on the scripted voiceover, and participants' conversations contained within these programmes, along with a thematic analysis of the situations contrived and/or purposively chosen by the television production companies to be portrayed in each episode of each programme. Specifically, each episode was viewed at least once. Recordings of the programmes were paused as necessary to record key phrases. Thus the research process was inductive and iterative. The data drove the analysis, and themes arose from an analysis of these phrases, scrutinising the data to find meaning as proposed by Braun and Clarke (2006) and Roberts and Pettigrew (2007), and undertaken based on a similar approach to that taken by Andriotis (2010) exploring the behaviours of young tourists in Greece through an analysis of print media rather than an analysis of the televisual media we focused upon.

As explained by Fereday and Muir-Cochrane (2006, p. 82), when employing an inductive thematic analysis, through carefully going over the data, it is possible to analyse the data to find "themes that emerge as being important to the description of the phenomenon"; these arise through "a form of pattern recognition within the data, where emerging themes become the categories for analysis". Thus, because our analysis was conducted inductively, we did not use any coding template that would otherwise be a feature of a deductive thematic analysis.

Findings

The television programmes included in our data set were mainly targeted at UK audiences, although the 1998 *GU* programme, while made by LWT was aired on the digital channel Sky One. All of the other programmes are now available to watch via the Internet, with many available on YouTube, which means that media audiences all around the world can see the destination image of Kavos as portrayed in these programmes.

GU

The focus of this series was to show young British tourism workers and holidaymakers indulging in alcohol-fuelled behaviour, focusing on "getting bollocksed" and "getting pissed", and showing extreme behaviours including drinking games, nudity, and sexual behaviour. Excerpts of the voiceover of this "documentary"-style reality programme in episodes focusing on Kavos include that "Kavos is Corfu's biggest party town, and the holiday season is getting into full swing" (Ep2) and that "In Kavos every night is party night" (Ep6).

SS&SP

In S1 and S2, the focus of the programmes was on groups of young people around 18 years of age who had just finished high school. S3 Ep6 featured 19-year-olds, but also on their first holiday without parents. The voiceover to S1 Ep4, set in Kavos, started as follows: "The first ever summer holiday with your mates, a rite of passage for every teenager, freedom, temptation and excess, all away from Mum and Dad". The voiceover goes on to state that "Corfu is full of history, beautiful beaches, sophisticated yacht clubs and Kavos, where every summer young Brits descend in their thousands, in search of sun, sea, holiday romance and cheap booze."

The voiceover for S2 Ep3 again focused on aspects such as "freedom" and "anything goes". "The freedom of your first ever holiday with your mates, where everyone's up for it and anything goes, cheap booze, dirty dancing, and that all-important independence away from Mum and Dad … . Coming up it's sun, sex and seriously bad behaviour in Kavos … party playground for young British tourists … the 24 hour party resort where everyone's up for it and anything goes" (S2 Ep3). This episode then showed a group of young lads setting off a fire extinguisher in their apartment on their first day's arrival, and the next day being informed of the cost to deal with the damage (€560) for what they termed a "prank", then heading off to the beach with cases of beer to 'phone parents asking for more money to be wired over to see them through the rest of their stay. On a night out one of the lads comments that "the girls out here are easy, they're just giving it away for free".

Both episodes of S1 and S2 that focused on Kavos each showed young holidaymakers requiring the attention of a doctor for drink-related injuries. S3 Ep6 showed an ambulance being called for one holidaymaker who was vomiting while lying in the street, and then passed out in the street through drink, and taken back to the medical emergency centre for treatment.

S3 Ep6 was introduced by the same initial voiceover as S2, focusing on the freedom of such a holiday with friends rather than with parents, but this season also stressed that the episode would take place "in the carnage and chaos of Kavos", which is a much stronger and more pejorative phrase than used in previous series set in the same destination. The introductory voiceover goes on to say that "The beautiful Greek island of Corfu is the Jewel of the Ionian Sea. It's also home to party town Kavos, rammed with bars and nightclubs, thousands of hedonists descend on its infamous Strip every week, looking for cheap booze, and even cheaper thrills."

The young British tourists the destination attracts are not portrayed in a particularly positive light. In the coach on the way to Kavos from the airport, the 19-year -old Essex girl subject of this episode asks her friends, "will there be, like, fish in the sea?", she is later seen vomiting in the street on the Strip after drinking. The lads' group were caught by hotel security urinating on an outside wall of the hotel on their way out to the Strip after drinking two bottles of spirits between four of them in their rooms. On another day, one of the group is seen throwing the towel he had used to clean up his vomit from the side of his bed onto another of the hotel's balconies.

Drinking games also featured in these episodes, and included a hollowed out watermelon filled with 28 shots, drunk by the group members from long straws, a holidaymaker sitting in a chair at a bar and having alcohol poured into his mouth via a large funnel, and large plastic syringes of cocktails being shot straight into holidaymakers' mouths. The Kavos *Booze Cruise* was featured in each episode of this programme that featured the destination. In S2 audiences also saw one of the holidaymakers ingesting laughing gas from a balloon.

C:TOTI

Although this programme is presented more as a documentary than a reality television programme, as has already been noted, it is becoming very difficult to tell exactly where the boundaries are between these two genres. In the opening voiceover the audience is told the following:

> This is Corfu, a small Greek island in the heart of the Med. In the winter it's a quiet traditional place, but in summer it's open for business. Corfu welcomes half a million Brits every year, and just like Britain it's a land of extremes … at one end is an exclusive enclave with some of the best luxury villas in the world. At the other is a party paradise where anything goes, then there's everything in between.

The focus of this series was ostensibly the Greek economic crisis and the effect of this crisis on tourism and the 10,000 or so expats who live on the island, noting that many of these expats also work in the tourism industry.

Kavos features from Eps 4–8 telling the audience that "most teenagers that arrive head South to the party town of Kavos, renowned for its nightlife". Ep4 showed Kavos early in the season highlighting a British workers' difficulty in earning money as a Public Relations (PR) representative (a job that involves encouraging tourists to go to a particular establishment). Similar to *SS&SP*, Ep5 of C:TOTI shows the "school leavers revelling in their escape from the family holiday". In Ep6, the content featuring Kavos focused on watersports, entertainment workers, large nightclubs, the Kavos drink culture, and young tourists' poor money management. "The first package holiday rolled into Kavos in 1982, and since then it's been a town of excess. For all the freedom on offer there's also a dark underbelly to the town." The content of this programme shows a worker in a large hotel in Kavos, seeing and commenting on the state in which young holidaymakers can leave their rooms.

WHIC

The introductory voiceover to the first season claims that

> Every summer the Greek resort of Kavos attracts a hundred thousand young Brits in the hope of having the time of their lives … here they escape their parents and the daily grind to party to the point of getting hospitalised. In this series holidaymakers tell us why they're flocking here in their thousands … with intimate access to the resort workers … the tourists and the locals that pick up the pieces … prepare yourselves for the complete uncensored story about life in a very modern paradise.

Each episode starts with straight-to-camera talking-head comments by those featured in the episode, including comments such as: "Everyone's wearing T-shirts saying 'Kavos fucks Ibiza'".

S1 Ep1: "In tonight's episode, sex and the fine art of courtship Kavos-style … life's consequences are felt in the resort where anything goes and the holidaymakers whose sexual injuries have the doctors baffled."

> "Until 30 years ago, Kavos on the Greek island of Corfu was just a sleepy fishing village. Then the first tourists arrived … the promise of sun, sea and sex lures up to 10,000 young Brits a week from all over the United Kingdom and the tourist trade does its best to make sure that they don't go home disappointed.

As with C:TOTI, this programme also features the Booze Cruise "a rite of passage for new arrivals", the programme shows drink games and sex simulation games. Within the first six minutes of the opening of S1 Ep1, two young female holidaymakers are seen to say to the camera "there's just no limits because it's Kavos", "there's no rules". In this episode male holidaymakers stated that their personal grooming regimes had been influenced by other reality television shows, namely *The Only Way is Essex* and *Geordie Shore*.

In this programme, the "infamous", "notorious" Strip is referred to as "a mile-long partyzone with over 60 bars, restaurants and clubs". Holidaymakers refer to seeing others having sex inside clubs, and in the street, one refers to a night where he had a "threesome on the beach".

The watermelon filled with seven shots of alcohol per person featured again as it did in SS&SP, as did the laughing gas that tourists are encouraged to inhale. There are many doctor's surgeries and clinics in the resort. One featured in this programme, the Kavos Emergency Centre, "treats over a thousand young Brits every season".

S1 Ep2 focuses on the drinking culture:

> In tonight's episode booze, Kavos-style, we meet the young holidaymakers who come here to party hard on the infamous Strip ... where the bars and clubs compete to get their customers as drunk as possible ... but the heady mix of limitless alcohol, no parents and no rules can have serious consequences".

Kavos Emergencies is described as charging €70 per visit, with the staff perceiving Kavos as being like a "warzone". Holidaymakers and doctors are seen discussing their beliefs that their drinks may have been spiked. The episode also featured people receiving medical treatment for falls, cuts, and injuries through fights and through random acts of violence, and features the cheaply priced "Headfucker" pint-sized cocktail, "often followed by a laughing gas chaser".

The programme does not only focus on the behaviour of holidaymakers, audiences are also informed that some of the younger Kavos workers are also abroad without their parents for the first time. As the voiceover regarding their behaviour and the aftermath of the wild nights out state: "It's a clash of cultures and generations."

S1 Ep3 then looks at the behaviour of the young workers, "the British youngsters who spend a summer working here", which the workers refer to as "getting laid and getting paid". In this episode, the focus was particularly on the bar PR workers, and considers the low pay and cost of medical treatment, and the cost of not being able to go to work if individuals are injured. This episode also shows tattooists tattooing drunk holidaymakers.

S2 similarly focuses on negative portrayals of the young British tourists the resort attracts: "What language do they speak here, is it Greekan, I think?"

S2 Ep1 focuses on exploring "the unique drinking culture of Kavos and its consequences", the introductory voiceover again focusing on what has happened to this destination since the influx of the type of tourist it attracts: "Kavos, once a sleepy fishing village on the south coast of Corfu, until the Brits arrived, transforming it into a Mecca for young holidaymakers intent on losing control". One of the young holidaymakers shown in this episode on his first ever lads holiday, away from parents sees the holiday as offering "such a sense of freedom, no restrictions".

S2 Ep2 then focuses on "the girls who come to Kavos to party hard".

S2 Ep3 focuses on those who have come to Kavos for the first time, for their first parent-free holiday away with their friends. However, unlike SS&SP, there is no final confrontation with parents who have indeed followed their offspring to the resort and witnessed their behaviour.

Summary

In summary, it can be seen that three main themes arose from our qualitative thematic analysis of each episode of these four reality television programmes, and these themes were common across all of the programmes. A holiday to Kavos, where a young person experiences the freedom of not being accompanied by parents, but instead is accompanied by friends of his or her own age is seen as a *right of passage*. It is this freedom of no longer being under the watchful eye of parents that facilitates the young British holidaymaker to indulge in risky behaviours and excessive consumption while adopting the view (gained from their own viewing of such programmes) that in Kavos, *anything goes*. Yet it is not only the young British holidaymaker that is seen behaving badly in these programmes, the young tourism workers may also be abroad on their own for the first time, and similarly indulge in the temptations and excesses of the resort, particularly casual sex, which they refer to as "getting laid and getting paid".

Right of passage

All four of these reality television programmes focused on typical Kavos tourists as "school leavers revelling in their escape from the family holiday" (C:TOTI, Ep5), arriving in large numbers: "10,000 young Brits a week from all over the United Kingdom" (WHIK, S1 Ep1). Activities such as the Kavos Booze Cruise are also seen in-resort as "a rite of passage for new arrivals" (WHIK, S1 Ep1).

Anything goes

The behaviour of these young British holidaymakers is typified as "no rules" and "anything goes". Most of the focus of the programmes was on night-time activities which mainly focus on the consumption of cheap alcohol (heavily featuring shots and cocktails in bars and nightclubs); intentional drug taking (particularly the inhaling of laughing gas through a balloon); unintentional drug taking (drinks having been "spiked");

drinking games (including those that simulate sex acts); and sexual activity; with a limited focus on daytime activities such as the Kavos Booze Cruise, watersport activities, go-kart racing, Zorbing and so on. The media also showed young people choosing to get tattooed while in Kavos, and tattooists undertaking this even when their clients were intoxicated. Other than organised trips that take place on the water, or in visits to the Aqualand waterpark, most of the tourists' bad behaviour is seen to take place on the "infamous", "notorious" Strip, "a mile-long partyzone with over 60 bars, restaurants and clubs" (WHIK, S1 Ep1). Tourists are also shown using local emergency medical services often for drink-related issues (a big feature of each episode of WHIK). One issue that was touched upon in WHIK is the cost of such medical treatment (€70 per visit before paying for any additional treatment).

Getting laid and getting paid

The tourism workers are also seen indulging in similar bad behaviours as the young tourists, focusing on alcohol consumption, sexual activity, and the need to use the services of medical facilities. WHIK (S1 Ep 3) showed the behaviour of the young workers, "the British youngsters who spend a summer working here", which the workers refer to as "getting laid and getting paid".

Discussion

The image of Kavos as a destination, as shown in these reality television programmes focuses, in the main, on the young British holidaymakers, abroad with their friends on their first holiday away from their parents. These tourists are seen to be indulging in the Kavos holiday experience as a means of escape from their everyday lives (Williams & Lew, 2015), acting in ways that they would otherwise not have the time to do, or have too much parental supervision to do (Andriotis, 2010; Wang, 1999). However, the sensationalised nature of this type of television programme focuses on only a few of the 10,000 young holidaymakers who visit the resort every week during the height of the season. Thus, how much the behaviour of the young people shown in these programmes is indeed a departure from their normal behaviour (Williams & Lew, 2015) is difficult to assess, as is how typical this behaviour is for the majority of Kavos tourists.

In the main, these programmes neglected to consider Kavos as a tourism destination by the families and older couples who holiday in the resort in the shoulder months. The main exception to this was C:TOTI, and this programme appeared to be leaning more towards documentary than "fly-on-the-wall" reality television of the more sensationalist type, but it is recognised that the boundaries between these genres are becoming blurred (Dunn, 2005; Nabi et al., 2003) and it is becoming increasingly difficult for audiences to evaluate which media portrayals are indeed authentic and which are simulated to increase entertainment value (Rose & Wood, 2005).

Thus these programmes were highly unbalanced in their portrayal of the changing patterns of tourism in the resort of Kavos over different periods of the year, and by the various different groups of actors in the destination conceptualisation process. A focus on specific types and patterns of particularly bad tourist behaviour could lead an audience to believe this particular aspect of behaviour is not only typical, but is the only type of tourist behaviour, thus leading to a skewed image of the destination that is not only portrayed unfavourably, but that is very much at odds with the reality of the place, at least for nine months of the year. While, on the one hand, this may lead the moral majority to perceive the destination image of Kavos as negative, and thus lead them to avoid holidaying there (Iwashita, 2006; Kim & Richardson, 2003; Lawson & Baud-Bovy, 1977), on the other hand, it may strengthen a positive perception of the destination image of Kavos as a place for potential tourists who want exactly the type of holiday portrayed in these programmes (Iwashita, 2003).

As far as its portrayal in these reality television programmes is concerned, the entire resort of Kavos has almost been reduced to the space of a one-mile Strip of bars and nightclubs. There are far more complexities to the place of Kavos than shown in television reality programmes, yet they continue to focus on the behaviour of some young holidaymakers who believe it is perfectly acceptable to trash their accommodation because they have not only seen it happen on television, but because such programmes continue to strengthen the belief in Kavos as a destination with no rules and where anything goes.

Conclusion

Our focus on the way media can influence the image of a destination in negative ways contributes to a very underdeveloped area in the literature on media-induced tourism in general (Beeton, 2005; Busby & Klug, 2001; Hudson & Ritchie, 2006; Skinner, 2016; Spears et al., 2013), and on the effect of reality television in particular. This therefore is our key theoretical contribution, to continue and further contribute to the development of the theoretical understanding of the relationship between reality television as a form of popular culture, and destination image, particularly with regard to portrayals of popular tourist destinations that have so far been lacking in the literature.

Unlike other forms of televisual media, reality television tends to be sensationalist and courts controversy. However, unlike scripted fictional drama that does not purport to show the "real", or documentary that does claim to portray the "real", while to at least some extent disguising editorial input, reality television purports to display reality, yet while it is often staged, and controversial behaviours are encouraged by the production team, such television is perceived as reality by its audience. Reality television also tends, at least to some extent, to portray its participants and their activities in an unfavourable light. In the case of reality television programmes such as these focusing on tourism in Kavos, portrayals of tourists behaving badly not only has the potential to affect audience perceptions of the tourists themselves, but also has the potential to affect audience perceptions of the destination that provides the location for their activities. Thus, reality television does appear to display characteristics not associated with other visual media associated with tourism, making it a suitable medium of separate academic enquiry. The phenomenon of reality television has to date been under-researched. This paper makes the first explorative small-scale qualitative enquiry into this phenomenon.

It is worth noting the limitations of this study. We recognise that highly qualitative research such as this will always draw negative claims regarding its subjectivity, and acknowledge with no little degree of frustration that "there is still … a lack of consensus on how to establish the legitimacy of visual tourism studies that draw on qualitative approaches" (Chambers, 2012, p. 39). However, we also believe that this research has value and will justify its legitimacy based on two main arguments: firstly, that we are researching a relatively new and under-researched phenomenon, thus it is explorative and such small-scale qualitative methods can be seen to be appropriate to exploratory enquiry; and secondly, that we are researching a phenomenon that is already highly subjective, and thus a subjective approach to its analysis could be seen to be more consistent than making an attempt to objectively analyse such a phenomenon from a positivist paradigm. Indeed, "classic attempts to take subjective data and organize them into some sort of systematic social scientific account have been seriously challenged on the grounds that they are imposing some kind of hierarchical order on the accounts of the participants themselves" (Gilhespy & Harris, 2011, pp. 63–64).

Our research also focused on only one destination and thus the generalisability of findings may be limited; however, the visual media portrayal of other such party destinations is very similar. Moreover, each of these party destinations is located in a place that offers many other types of tourism product to different target markets, each of which may find particular destination spaces contested by a range of stakeholders.

If the recommendation in the literature on negative place identities is to be followed, Kavos should do everything it can to overcome the potential for the formation and perpetuation of a negative image (Bastakis et al., 2004; Kamenidou et al., 2013). The perception of the place in a "negative" light is based on consumption behaviour by a specific target market segment for only three months of the year. The rest of the time it reverts to the normal way of life for many Greek coastal island resorts. Those local people that own businesses in the resort rely upon tourism for their income, and appear to enjoy the benefits they reap. Should Kavos, and other party resorts, move to a more up-market demographic, this would simply push such types of tourism to other places, and require an amount of infrastructure investment that can be ill afforded.

What really happens in Kavos is not only what is portrayed in the media. The reality televisual media portrayal of the destination image of Kavos focuses on only one narrow segment of young British high-season tourists, and is therefore highly unbalanced and not only neglects the changing patterns of tourism during different periods of the year, but also neglects various different groups of stakeholders in the destination image formation process, leading to a skewed perceived image of the destination that is not only negative, but very much at odds with the reality of the place, at least for nine months of the year. This leads to some tourists staying away because either they believe nothing is open to serve them in the shoulder months, or because they do not perceive they "fit" with the stereotypical tourist shown in these programmes. Many of the tourist behaviours shown in these programmes are classed as risky, and can be perceived as negative by those not belonging to the young partying target market.

Televisual media content is known to be able to affect destination image, and such content is now more easily accessible across a range of platforms, including the Internet, which itself is being more heavily used by potential tourists to make decisions about destinations. Reality television programmes are increasingly considered by many audiences to help in their quest for authenticity, because, even when known to be staged in some way, such programmes are indeed deemed to offer a more real portrayal of the people and places they inhabit than other more traditional forms of fictionalised drama or scripted documentary.

This type of visual media portrayal about places is one that should be more thoroughly researched. Further research could consider other destinations in order to ascertain if different types of reality television portrayals project different destination images. Future studies could also fruitfully focus on the way a wider range of destination images portrayed in reality television programmes are perceived by various audiences, including other online and offline media, and by potential tourists themselves.

Disclosure statement

No potential conflict of interest was reported by the authors.

ORCID

Heather Skinner http://orcid.org/0000-0002-6505-8073

References

Alonso, I., & Bea, E. (2012). A tentative model to measure city brands on the Internet. *Place Branding and Public Diplomacy, 80*(4), 311–328. doi:10.1057/pb.2012.22

Andriotis, K. (2010). Brits behaving badly: Template analysis of newspaper content. *International Journal of Tourism Anthropology, 1*(1), 15–34. doi:10.1504/IJTA.2010.036844

Bastakis, C., Buhalis, D., & Butler, R. (2004). The perception of small and medium sized tourism accommodation providers on the impacts of the tour operators' power in Eastern Mediterranean. *Tourism Management, 25*(2), 151–170. doi:10.1016/S0261-5177(03)00098-0

Beeton, S. (2005). *Film induced tourism*. Clevedon: Channel View Publications.

Blackwell, R. D., Miniard, P. W., & Engel., J. F. (2005). *Consumer behaviour* (10th ed.). Cincinatti, OH: South Western College Publishing.

Braun, V., & Clarke, V. (2006). Using thematic analysis in psychology. *Qualitative Research in Psychology, 3*(2), 77–101. doi:10.1191/1478088706qp063oa

Brayshaw, D. (1995). Negative publicity about tourism destinations - a Florida case study. *Travel and Tourism Analyst, 5*, 62–71.

Busby, G., & Klug, J. (2001). Movie-induced tourism: The challenge of measurement and other issues. *Journal of Vacation Marketing, 7*(4), 316–332. doi:10.1177/135676670100700403

Chambers, D. (2012). The [in]discipline of visual tourism research. In T. Rakić & D. Chambers (Eds.), *An introduction to visual research methods in tourism*. London: Routledge.

Crompton, J. L. (1979). An assessment of the image of Mexico as a vacation destination and the influence of geographical location upon that image. *Journal of Travel Research, 17*(4), 18–23. doi:10.1177/004728757901700404

Davidson, L., & Skinner, H. (2010). I spy with my little eye: A comparison of manual versus computer aided analysis of data gathered by projective techniques. *Qualitative Marketing Research: An International Journal, 13*(4), 441–459. doi:10.1108/13522751011078845

Dunn, D. (2005). We are *not* here to make a film about Italy, we *are* here to make a film about ME ...: British television holiday programmes representations' of the tourist destination. In D. Crouch, R. Jackson, & F. Thompson (Eds.), *The media and the tourist imagination: converging cultures (contemporary geographies of leisure, tourism and mobility)* (pp. 154–169). Abingdon: Routledge.

Echtner, C. M., & Ritchie, J. R. (1991). The meaning and measurement of destination image. *Journal of Tourism Studies, 2* (2), 2–12.

Fakeye, P. C., & Crompton, J. L. (1991). Image differences between prospective, firsttime, and repeat visitors to the Lower Rio Grande Valley. *Journal of Travel Research, 30*(2), 10–16. doi:10.1177/004728759103000202

Fereday, J., & Muir-Cochrane, E. (2006). Demonstrating rigor using thematic analysis: A hybrid approach of inductive and deductive coding and theme development. *International Journal of Qualitative Methods, 5*(1), 80-92.

Gartner, W. (1993). Image formation process. *Journal of Travel & Tourism Marketing, 2*(2–3), 191–216. doi:10.1300/J073v02n02_12

Gilhespy, I., & Harris, D. (2011). Representing 'the real': Realism and visual culture in tourism, leisure and ethnography. *Journal of Tourism Consumption and Practice, 3*(1), 51–68.

Gunn, C. A. (1972). *Vacationscape: Designing tourist regions*. Austin: University of Texas.

Hanson, R. K. (1992). Thematic analysis of daily events as a method of personality assessment. *Journal of Personality Assessment, 58* (3), 606–620. doi:10.1207/s15327752jpa5803_14

Hudson, S., & Ritchie, B. J. R. (2006). Promoting destinations via film tourism: An empirical identification of supporting marketing initiatives. *Journal of Travel Research, 44*(4), 387–396. doi:10.1177/0047287506286720

Iwashita, C. (2003). Media construction of britain as a destination for Japanese tourists: Social constructionism and tourism. *Tourism and Hospitality Research, 4*(4), 331–340. doi:10.1177/146735840300400406

Iwashita, C. (2006). Media representation of the UK as a destination for Japanese tourists: Popular culture and tourism. *Tourist Studies, 6*(1), 59–77. doi:10.1177/1468797606071477

Kamenidou, I. C., Mamalis, S. A., Kokkinis, G., & Geranis, C. (2013). Image components of nightlife-clubbing destinations. *Tourismos, 8*(3), 99–111.

Karpovich, A. I. (2010). Theoretical approaches to film-motivated tourism. *Tourism and Hospitality Planning & Development, 7*(1), 7–20. doi:10.1080/14790530903522580

Kim, H., & Richardson, S. L. (2003). Motion picture impacts on destination images. *Annals of Tourism Research, 30*(1), 216–237. doi:10.1016/S0160-7383(02)00062-2

Lawson, F., & Baud-Bovy, M. (1977). *Tourism and recreational development*. London: Architectural Press.

Lichrou, M., O'Malley, L., & Patterson, M. (2008). Place-product or place narrative(s)? Perspectives in the marketing of tourism destinations. *Journal of Strategic Marketing, 16*(1), 27–39. doi:10.1080/09652540701794429

McElroy, R. (2011). Putting the landmark back into television: Producing place and cultural value in Cardiff. *Place Branding and Public Diplomacy, 7*(3), 175–184. doi:10.1057/pb.2011.20

Morgan, N., & Pritchard, A. (1998). *Tourism promotion and power, creating images, creating identities*. Chichester: Wiley.

Nabi, R. L., Biely, E. N., Morgan, S. J., & Stitt, C. R. (2003). Reality-based television programming and the psychology of its appeal. *Media Psychology*, *5*(4), 303–330. doi:10.1207/S1532785XMEP0504_01

O'Sullvan, K. (2013). Channel 4's what happens in Kavos is their idea of entertainment - gormless guys drinking urine and vomit. *The Mirror* [Internet]. Retrieved December 15, 2013, from http://www.mirror.co.uk/tv/tv-reviews/channel-4s-what-happens-kavos-2870649

Rakić, T., & Chambers, D. (Eds.). (2012). *An introduction to visual research methods in tourism*. London: Routledge.

Roberts, M., & Pettigrew, S. (2007). A thematic analysis of children's food advertising. *International Journal of Advertising*, *26*(3), 357–367.

Rose, R. L., & Wood, S. L. (2005). Paradox and the consumption of authenticity through reality television. *Journal of Consumer Research*, *32*(2), 284–296. doi:10.1086/432238

Schwartz, D. (1989). Visual ethnography: Using photography in qualitative research. *Qualitative Sociology*, *12*(2), 119–154. doi:10.1007/BF00988995

Shagrir, O. K. (2015). Unveiling television's apparatus on screen as a 'para-interactive' strategy. *Media, Culture & Society*, *37*(5), 737–752. doi:10.1177/0163443715577245

Skinner, H. (2011). An archaeological excavation into the fields of place marketing and place branding. In A. Patterson & S. Oakes (Eds.), *Proceedings of the academy of marketing conference 2011: Marketing fields forever*. Liverpool: Academy of Marketing.

Skinner, H. (2016). What's occurring? Barry since Gavin & Stacey. *International Journal of Tourism Research*, *18*(3), 251–259. doi:10.1002/jtr.v18.3

Spears, D. L., Josiam, D. M., Kinley, T., & Pookulangara, S. (2013). Tourist see tourist do: The influence of hollywood movies and television on tourism motivation and activity behavior. *Hospitality Review*, *30*(1), Article 4. Retrieved from http://digitalcommons.fiu.edu/hospitalityreview/vol30/iss1/4

Tasci, A. D. A., & Gartner, W. C. (2007). Destination image and its functional relationships. *Journal of Travel Research*, *45*(4), 413–425. doi:10.1177/0047287507299569

Volpe, L. (2013). Boozy Brits abroad: What happens in Kavos should stay in Kavos: The drunken antics of Brits abroad are appalling – and we're only encouraging them. *The Independent* [Internet]. Retrieved December 15, 2013, from http://www.independent.co.uk/voices/comment/boozy-brits-abroad-what-happens-in-kavos-should-stay-in-kavos-8734229.html

Wang, N. (1999). Rethinking authenticity in tourism experience. *Annals of Tourism Research*, *26*(2), 349–370. doi:10.1016/S0160-7383(98)00103-0

Williams, S., & Lew, A. A. (2015). *Tourism geography: Critical understandings of place, space and experience* (3rd ed.). Abingdon: Routledge.

Winston, B. (2008). *Claiming the real: Documentary: Grierson and beyond*. Basingstoke: Palgrave Macmillan (on behalf of the British Film Institute).

(Re)negotiating authenticity through virtual travel: a case study of *Law of the Jungle*, a Korean reality travel program

Sohye Kim, Sanghun Park, Carla A. Santos and Kimberly J. Shinew

ABSTRACT

The current study investigates tourists' discourse regarding authenticity in the postmodern tourism context by employing the intertwined concepts of authenticity and gaze to examine *Law of the Jungle*, a Korean reality travel program. Findings from our thematic content analysis of data suggest that the program provides viewers with a dual structure through which to view and negotiate the tourism setting of the show, and that the authenticity of the program is perceived from three main perspectives: objective, subjective, and postmodernist. Findings are contextualized within extant literature on authenticity, orientalism, the tourist gaze, and the second gaze.

Introduction

> The program's a complete lie! What's this? No less than a fictional drama. A real documentary variety show? Fxxx. Does the camera catch a pure reaction by forcing [the actors] to eat something they don't want to and releasing animals around them, huh? Why don't you go to NY and hunt squirrels next time, xxx? Choosing a country where they want to travel and buying a ton of beer for $1000 all night, [they] now go drinking beer in the pub every night! Got a best program award, what a xxx!

On February 5 2013, the owner of a Korean talent agency, Sang-You Kim, posted a message on Facebook that showed up and brought to light the fabrication of the popular reality show, *Law of the Jungle*. Kim once worked on the show as a crewmember tasked with caretaking for a cast member. According to Kim, while show participants were represented in the program as surviving in a jungle, they were in fact partying all night at a pub near the hotel where they stayed. Although he deleted the posting immediately after receiving considerable attention from viewers and the media, and blamed being drunk as an excuse for his disclosure, this incident led audiences to perceive a different reality beyond the show's official representation and provoked a fierce public discourse regarding the authenticity of the "reality travel" show.

This incident provided an enticing case for exploring authenticity in the context of postmodern tourism, given that the act of watching travel shows enables viewers to be virtual travelers and erodes boundaries between reality and fantasy. This erosion has been given substantial attention by tourism scholars who to date have largely focused on the meanings of tourists' experiences that are negotiated based on their own subjective interpretations (Cohen, 2008; Lash & Urry, 1994; Uriely, 2005). Contrary to the interest at the theoretical level, little attention has been devoted to watching reality travel shows as a practice of postmodern tourism, despite the fact that the programs are not simple entertainment but serve as a venue for consuming authenticity in a given cultural context.

Seeking authenticity is a reconciliation process between personal narratives and the socially systemized norms and values implanted in the toured objects by the media. Entertainment media provide audiences, regardless of their physical locations, with an extended field of view of the world through the television screen (Strain, 2003). Viewers as "armchair" tourists (Strain, 2003; Wollen, 1991) seek objects to gaze upon such as the representations of tourism landscapes that they internalized through guidebooks (Chang & Holt, 1991), postcard images and brochures (Markwick, 2001; Molina & Esteban, 2006), travel writing (Santos, 2004), and travel programs (Fürsich, 2002). These pre-established expectations of tourism destinations, local people, and their cultures reproduce socially constructed notions of authenticity (MacCannell, 1973). However, despite such manufactured tourism contexts, viewers are still capable of constructing a subjective sense of authenticity when they recount their own original experiences away from the typical tourism narratives shaped by institutional and tourism industrial perspectives (Shaffer, 2004).

The concept of the *gaze* provides an interesting and appropriate lens to capture this tension. In particular, Urry's (1990) work added a sophisticated dimension to the concept of authenticity through *the tourist gaze*. Through the gaze, tourism landscapes and their representations are visually consumed beyond the physical destination, and the image of localities created by tourism industries systematically directs the first-hand experiences of tourists. In this context, a more nuanced form of authenticity is determined by the degree to which the first-hand experience confirms, negates, or negotiates the representations of the toured objects. Despite its extensive applicability to diverse tourism phenomena (see for example Maoz, 2006; Pritchard & Morgan, 2000; Wilson & Suraya, 2004; Zhihong, 2007), the adoption of this Foucauldian perspective has been limited in investigating a tourist experience. In addition to Urry's (1990) work on the gaze, MacCannell (2001) has argued the existence of resistance to the tourist gaze in the tourist's visual practices and referred to it as *the second gaze*. However, to date only a handful of scholars have discussed the diverse dimensions of gaze in relation to authenticity (see for example Kirschenblatt-Gimblett, 1998). When considering that the concepts of authenticity and gaze are intertwined with the sociocultural nature of tourism (Selby, 2003), and that entertainment reality programs provide viewers with armchair tourism experiences in which they encounter the paradoxical nature of authenticity (Rose & Wood, 2005), what is lacking is an exploration of how the viewers of entertainment media authenticate their virtual travel via gazing.

With this in mind, this study aims to investigate how spectators perceive the authenticity of a particular reality travel program through their practice of seeing. Specifically, this study examines the narrative structure of *Law of the Jungle* and then explores the viewers' discourse regarding the authenticity of the locations, hosts, cast members, and show producers. To do so, it analyzed the dual structure of the program: an entertainment-driven framework that drives viewers to seek vicarious satisfaction through cast members who are positioned and framed as travel agents, and to gaze upon the images on the screen as toured objects. We also explored the online discourse among the show's viewers that revealed the multifaceted nature of authenticity in this postmodern tourism setting. The research is guided by an attempt to answer the following research questions: (1) How does the program *Law of the Jungle* represent the toured objects?; and (2) How does the discourse employed by viewers of the show describe the authenticity of their virtual travel experience? Answering these questions on a practical level requires us to describe the features of the show as well as probe the theoretical and conceptual issues of authenticity and gaze.

Program description

Law of the Jungle is a popular South Korean reality travel program that has been broadcast since October 2011. The main narrative of the program is for participants to find their lost identity and challenge their new selves in remote, pristine, and primitive natural settings. The premise of the show is that Korean celebrities visit areas where the way of life is maintained, unchanged, uncivilized, and unrestrained (Echtner & Prasad, 2003). The program portrays the show participants' experiences as they explore a traditional way of life in remote, pristine, and natural destinations, such as Namibia and other African countries, Papua New Guinea, Vanuatu, Siberia (Russia), Madagascar, Galapagos Islands (Ecuador), Amazon rainforest, New Zealand, Micronesia, Thailand, and New Caledonia (French overseas territory). The travel program proposes to portray the cast members' experiences as real and authentic since they are exposed to and experience the traditions of local primitive tribes (e.g. Batak, Waorani, and Maori) through their immersion in the local settings and culture as they learn how to survive in nature.

This program has fascinated and attracted audiences in South Korea by employing powerful elements, such as travel, authenticity, adventure, and entertainment. However, the Facebook disclosure by Sang-You Kim of the manipulated aspects of this successful top-rated program initiated an intensive debate on its authenticity. Moreover, indisputable evidence reported by viewers revealed the manipulated setting of the program. For example, they pointed out that the various settings of the program were not remote destinations but rather popular tourist destinations, and that the people portrayed as locals were, in fact, workers engaged in the tourism industry.

In an effort to alleviate the mounting concerns, the broadcast station of the show (Seoul Broadcasting System), held a press conference in which Byung-Man Kim, the chief member of the show, pleaded with audiences, stating:

> We tried our best to truly experience the site, whether our visited places were tourist destinations or not. ... Regardless of the authenticity of the cultural settings and the indigenous people, we really experienced the locals' traditional culture. We did indeed live in the jungle. ... In fact, there couldn't be a 100% reality

show unless we got lost [in the jungle] (Hu, February 12 2013).

Despite his appeal, the program's ratings decreased from 17.88% (January–February 1, 2013) to 14.83% (February 8–December, 2013) after the disclosure (Table 1), and there still remained controversial issues regarding how, if at all, the program manipulated that which it represented. A significant number of comments and criticisms of the authenticity of the program appeared on online interfaces, thus providing an excellent opportunity to interpret the audience's idea of the meaning of authenticity in the postmodern tourism context of South Korea.

Literature review

Authenticity

The development of scholarly discourse on authenticity in tourism is in line with the historical evolution of the West from modernity to postmodernity. Modernism reduces the world with a unifying social order in which absolute dichotomous categories exist of women and men, Orient and Occident, subject and object, real and fake, and daily routines and extraordinary experiences. In the modernist perspective, identifying authenticity of toured objects is crucial for tourists in order to experience the "real". In this vein, Boorstin (1964, p. 255)classified mass tourism sites as pseudo-events and argued that, in the "age of contrivance" modern tourists, despite their efforts, are unable to experience authenticity in such settings because tourism spaces are manipulated to elicit fake experiences. Moreover, MacCannell (1973) applied Goffman's (1959) idea of front stage and back stage to the tourism context and emphasized the contrast between the front regions, which are inauthentic due to their artificial construction for tourism purposes, and the back regions, which are authentic because the locals use the space for their everyday lives.

These object-oriented approaches are questioned by social constructivists who argue that authenticity is not objectively inherent in a toured object but symbolically projected by tourists or tourism industries (Wang, 1999). Taylor (2001) criticized tourism for having appropriated the term to strengthen the value of the toured objects even though it is socially constructed as an ideological concept in Western modernism. In this context, the value of an object is a result of power struggles legitimated by those who have the right to authenticate its meaning (Bruner, 1994). Thus, social constructivists claim that this concept of authenticity needs to be re-examined for its constitutive nature that is negotiable (Cohen, 1988), context-oriented (Bruner, 1989), and power-related (Silver, 1993).

Postmodernism scholars further claimed the need to ignore toured objects and instead focus on the deconstruction of authenticity by touring subjects. Baudrillard (1983) argued that the modernist concept of authenticity has collapsed, or been subverted at least, by postmodern tourists whose experiences are simulacra of reality in which the sharp contrast between the real and the fake is blurred. Under this postmodern condition of the tourism context, geographical experiences become a form of simulacrum in which the boundary between the actual landscape and its representation is eroded (Harvey, 1990), and "post-tourists" are able to not only recognize an inauthentic tourism setting but also enjoy its fabrication (Urry, 1990). Lash and Urry (1994, p. 275) interpreted the experiences of post-tourists as "merely a series of games with multiple texts", and in turn, did not raise the issue of inauthenticity. Anchored in Baudrillard's (1983) work, Eco's (1986) idea of "hyper-reality" highlights the postmodernist infertility of authenticity in the tourism context: the "Disneyfied" world where we are embedded deconstructs the boundary between the copy and the original, and so it is meaningless to seek out reality-based authenticity as MacCannell (1973) did (Eco, 1986). Thus, deconstructing authenticity is inescapable given that the original is no longer used as a reference to the world of simulacrum in which the boundary between sign and reality and between fantasy and reality is blurred.

Within this feature of hyper-reality lies a paradox of the postmodern tourist: although tourists savor the manipulated scenes, the modernist value of the real is desired. This tendency manifests itself in the reality travel program that is the focus of this study, which is a representative case of postmodern tourism with its popularity and superfluity (Hartley, 1999; Patino, Kaltcheva, & Smith, 2011). Rose and Wood (2005, p.

Table 1. Annual national program rating of *Law of the Jungle* from 2011 to 2016.

			2013					
Year	2011 (Oct–Dec)	2012	Jan–Feb 1	Feb 8–Dec	2014	2015	2016 (Jan–July)	Total
Program rating (%)	11.89	15.59	17.88	14.83	13.21	12.40	12.93	13.67

Source: Drafted by the authors based on information collected from AGB Nielsen Korea Media Research (www.nielsenkorea.co.kr).

286) noted that the success of reality-based shows is "a reflection of the modernist's nostalgia for authenticity" among viewers. Even though "virtual travel" blurs the margins between the actual travel and its images (Urry, 2002, p. 144), the vision of audiences is indirectly and directly regulated through media (Crary, 1990). Thus, while the postmodern collapse provides tourists with a panoramic experience between travel and its representation (Baudrillard, 1983), vision as an act of seeing could still be subject to a certain discipline of specific institutions which implant the modernist values via the tourist gaze (Crary, 1990; Friedberg, 1993).

The gaze and tourism

The tourist gaze is defined as the tourists' visual consumption of toured objects in "socially organised and systematised" ways (Urry, 1990, p. 1). MacCannell (2011) described three contributions of this term to the exploration of the contemporary tourism phenomena: to veer attention from tourism service providers to tourists; to expand the range of research topics from economic and marketing values to social-psychological concerns; and to erode the binary between reality and fantasy. Despite its substantive significance, criticisms of the tourist gaze have been raised in two strands. First, this vision-oriented concept reduces the arena of tourism to sightseeing, and in turn, attenuates the importance of the corporeal and physical aspects of tourism performances. Second, Urry (1990) does not really consider the possibility that what he describes is *a* tourist gaze of the visionscapes.

An ontological blind spot in the concept is identified by tourism scholars who emphasize the somatic forms of experience (Crouch, Aronsson, & Wahlstrom, 2001; Edensor, 2001; Franklin & Crang, 2001; Perkins & Thorns, 2001). For them, Urry's (1990) simplified approach to "the body as an object" is incapable of accounting for the multisensory nature of tourist experiences that are relatively free from vision-oriented knowledge (Crouch, 2000, p. 68). In an effort to overcome the limited interpretation of the term, recent studies have employed it in a variety of multisensory, somatic, and non-representable tourism contexts ranging from adventure (see for example Cloke & Perkins, 1998) to food (see for example Everett, 2009).

Admitting the vision-exclusive aspect of the term, Urry (2002) later elucidated that the literal denotation of "gaze" never implies one-dimensionality (Franklin & Crang, 2001) and refined the multiplicity of the gaze in relation to embodied experiences by mentioning that "the body senses as it *moves*" (Urry, 2002, p. 153). He further investigated the similarities between the gaze-centered approach and the performance-oriented paradigm in order to expand the coverage of the term from the static and passive way of sightseeing to a broad spectrum of signs, practices, and experiences (Urry & Larsen, 2011). In his responsive argument, an interesting point is that Urry (2002; Urry & Larsen, 2011) consistently considers visual perception as the master sense, so it is privileged in tourism practices.

His tenacious hold on ocularcentrism, however, is criticized for his inattentiveness to possible layers within the field of vision. Zhihong (2007, p. 253) argued that the gaze is not unidimensional because "It presents a chance for domestic tourists to gaze at foreign tourists and consume a foreign atmosphere". Lee (2001) pointed out that Urry's (1990) approach is based on the Eurocentric ideas of nostalgia and romanticism. Jordan and Aitchison (2008) and Maoz (2006) emphasized the need to explore mutual gazes between the tourist and the locals, both of which are identically important to understanding power generation.

MacCannell (2001, 2011) criticized Urry at the theoretical level for simply dividing the tourist experience into "seen" and "unseen" due to the dependence on Foucault's medical gaze. Probing the complex relationship between knowledge and power, Foucault (1963) argued that gaze is the most privileged and important sense of human perception and the medical gaze is tied to power in a discursive way such that the observations of doctors are delivered as the truth to patients and students. In Urry's (1990) words, the medical gaze is replaced by the touristic gaze with discourse organized by authorities ranging from knowledge-generating institutions to many professionals (Larsen, 2014). In this context, the visual practice of tourists is mediated through ideal representations and knowledge of landscapes shaped by the tourist gaze (Taylor, 1994; Urry, 1990).

For MacCannell (2001, p. 30), this closed system of looking is tricky in that tourists who remain seemingly free within the fixed structure in fact "cannot be free" to go beyond the order. He claimed that the tourist gaze in the deterministic interpretation of freedom only describes *a* gaze that echoes "the narcissistic needs of dull egos" and is incapable of explaining a change beyond the sociostructural frame (MacCannell, 2001, p. 26). Thus, anchored in Lacan (1981) psychoanalytical interpretation of gaze, MacCannell (2001) argued the second gaze that is "capable of recognizing the misrecognition that defines the tourist gaze" (MacCannell, 2001, p. 30). This notion enables him to interpret tourists who are not only "exhausted by the condition of its visibility" but also "resistant to the limitations of the tourist gaze" from which something is "concealed", "missing", and "hidden" (MacCannell, 2001, pp. 30–36).

What is being emphasized in his argument for the second gaze is the importance of freedom for the human subjects in order to approach the real: "Tourists are human beings. They must have free will. ... It [the second gaze] looks for the unexpected, not the extraordinary, objects and events that may open a window in structure, a chance to glimpse the *real*." (MacCannell, 2001, pp. 23, 36; italics added). MacCannell's (1973) interpretation of staged authenticity might provide a pertinent insight contributing to the comprehension of the real in the second gaze. For MacCannell (1973), unlike the front stage, the back stage is concealed from the view of tourists and is authentic due to its use by the locals for their daily lives. Common features between MacCannell's (2001) approach to the invisible and MacCannell's (1973) back stage are that while they are visibly or spatially "hidden," they nonetheless embrace an actual facet of the toured object. This metaphoric cohesion alludes to his focus on the real: it is not the truth of an object, but tourists' realization of the false premise that the eyes of tourists provide themselves with a pure, unmodified vision of the toured object regardless of its representation by the tourism industry (MacCannell, 1989, 2011). In this context, a tourist's reality is not necessarily the actual historical, objective truth; therefore, tourists who resist the signs manipulated by markers are able to continue searching for real meaning through the second gaze despite their conscious awareness that sightseeing is always "mediated".

The gaze: from tourism to media

Even though the search for the real in the second gaze encourages tourists to resist the tourist gaze (MacCannell, 2001), the camera objectifies the tourist gaze and recreates its power in accordance with the hegemony of vision (Bruner, 2005; Jay, 1993). In this context, the toured entity that is seized by the camera functions as a node of socially constructed representations in the power system. In media representations of tourism landscapes, however, the relationship between the toured object and the touring subject is not solely determined by the tourist gaze (Cohen, Nir, & Almagor, 1992). This is because tourists are self-reflective in the venue of tourism (MacCannell, 2001), and the images of others presented in the media provide them with the opportunity to reflectively perceive themselves through the representations (Said, 1978). Considerable attention has been devoted to the link between media and tourism by exploring how the representations of people, destinations, and cultures are described in diverse forms of visual media such as television and videos (Brown & Singhal, 1993), movies (Beeton, 2006; Riley & Van Doren, 1992), and brochures (Echtner & Prasad, 2003). These media technologies share a commonality with tourism in that they all replace so-called facts with representations (Beeton, 2005), and all the representations can be commodified and advertised (Teo & Yeoh, 2001).

This resemblance leads scholars in cinema studies to apply the tourist gaze in their field but with a different focus. Friedberg (1993) employed this term in connection with cinema and postmodernism to account for the commodification of a virtual gaze in which the spectator is able to travel via cameras and projectors that bring distant touristic landscapes to the screen. The mobility in cinema, therefore, excludes the movement of the audience's body, and the touristic gaze here serves as an apparatus to explore a simulated reality. This "virtuality" in the cinematic gaze underscores the postmodern condition of viewing that is free from the quest of the real (Strain, 1998, p. 150). Friedberg's interpretation of the tourist gaze separate from bodily experiences raises a question of how to reconcile the term of the cinematic perspective, which has no concern for corporeal space, and that of tourism, which traditionally tends to be more rooted in the real and physical movement of tourists.

A semiotic approach to tourism allows us to situate the cinema-related media experience as a virtual practice of tourism; it is the tourist who reduces the tourist attractions to signs of themselves, and it is impossible to perceive any environment that is unmediated (Culler, 1981). Given this, tourism aligns with many other forms of human experiences that are incapable of offering the human subject any direct interaction with the real. Producers of reality television programs and filmmakers use the illusion of the imaginary real to make viewers feel as if they were in the scene they are watching. The depiction of exotic others through the illusion of de-mediation is often judged "according to their capacity for extending this illusion of direct experience" (Strain, 1998, p. 154). The scene on the screen becomes authentic and original when the representations are perceived as fitting the image stored in the viewer's memory (Buzard, 1993). Thus, in the venue of virtual travel, spectators could call for and claim authenticity through gazing upon representations projected and embedded in the myth of direct experience.

Strain's (2003) gaze touches the mythical aspect of gaze by arguing the illusion of de-mediation. She applied Urry's (1990) semantic definition of the tourist gaze to cinema and television viewing as cultural practices given that the virtual mobility provided by film and television enables spectators to be "armchair tourists" without corporeal movement (Wollen, 1991, p. 191). Anchored in Friedberg's mobilized virtual gaze, she places emphasis on the tourist gaze with its virtual "mobility" in which viewers metaphorically travel in "the collapse between the deployment of this gaze in virtual and actual situation" (Strain, 2003, p. 279). This perceived freedom of movement is a cinematic illusion; while advanced entertainment media technologies provide viewers with more authentic and direct experience, this seemingly de-mediated experience is in fact enacted by their physically distanced position from the scene (Strain, 1998). This paradoxical viewing practice highlights the dual position of the tourist gaze; viewers simultaneously *absorb* the contents while *distancing* themselves from "being there" (Strain, 2003, p. viii). Considering the constant forces of push and pull in the spectator's vista, Strain (2003) argued that the tourist gaze functions as an operational tool broadly deployed in diverse cultural practices in terms of fetishizing the toured objects on screen.

With respect to the theoretical and conceptual approaches to the tourist gaze from Urry (1990) to Strain (2003), it seems natural that one expects accessibility to a certain form of authenticity through reality television programs. If the viewer's quest for authenticity is fulfilled by matching representations in the media with images shaped by the tourism industry, producers of television programs must provoke the tourist gaze by depicting and highlighting the exotic, extraordinary features of the localities that the industry has provided as tourism attractions. This argument is based on two assumptions. First, the way of gazing is universal. Thus, all viewers, regardless of their cultural backgrounds, interpret the representations of the toured objects in the same way. Second, media-mediated tourism is always situated in an uncritical context. Thus, spectators are induced to gaze uncritically at the representations on the screen. If touristic sightseeing through a television show fails to hold these two assumptions, the critical second gaze would occur in order to "glimpse the real" by distancing viewers from or immersing them in the scenes. The disclosure of the manipulated, staged tourism setting of *Law of the Jungle* is an interesting case that deviates from the above assumptions.

Given this, the current study analyzes the narrative structure of the program and the ensuing discourse among audiences regarding authenticity that is simulated by gazing upon the scenes of the show. It explores: (1) how the program *Law of the Jungle* represents the toured objects; and (2) how viewers of the show describe the authenticity of their virtual travel experience.

Method

Data collection

Data was derived from two sources: (1) an analysis of the narrative structure of the reality television show, *Law of the Jungle*, and (2) textual data derived from viewers' comments about the authenticity of the program on two online news forums. Since television functions as a storyteller that reflects values and both implicitly and explicitly identifies a dominant ideology (Fiske & Hartley, 1978), investigating the structure of the reality television show allows us to begin to understand how the structural nature of televised shows manipulates scenes on the screen so that viewers feel object-oriented authenticity beyond the television screen. In this context, we examined the interactions between the producer, the actors, and the viewers in constructing authenticity through the narrative structure of *Law of the Jungle*. This activity provided a context for understanding the gazing structure of the show.

Viewers' comments, our second data source, were collected from publicly available online comments about articles posted on online news forums. Although the Facebook message by Sang-You Kim initially provoked viewers to discuss their thoughts regarding authenticity of the program, it was specifically through online news forums that they actively displayed their opinions about the issue in depth. This reflects a cultural tendency in Korea to utilize news forums housed on online portals as a more popular practice for posting comments on social issues compared with that of other social media platforms such as Facebook. Thus, with respect to this cultural aspect, we considered online articles and their corresponding comments posted on online portals for data collection. Two representative portals (www.daum.net and www.naver.com) were selected for two primary reasons. First, due to the popularity of the websites, we could obtain a significant number of comments in which users fiercely discussed the authenticity of the reality program. Second, there was no danger of violating anyone's privacy with textual data obtained from the two sites, but they did provide trustworthy comments of the viewers.

Using the keywords *Law of the Jungle* and "authenticity", a total of 697 news articles (128 on Daum and 569 on Naver) published during February 2013, soon after the disclosure of manipulation by the program, were identified. Among these articles, 52 were specifically selected for having the largest number of comments attached, ranging from 466 to 10,124. Even though the articles were initially published in February, the times when users made comments vary between February 2013 and May 2014. Consequently, a total of 78,541 comments were collected and analyzed. All of the collected text data were translated from Korean into English and then backtranslated into Korean to ensure accuracy in translation. This process was carried out by the first and second authors, both bilingual speakers of English and Korean.

Data analysis

Through a process of integration and triangulation, the text data were analyzed and interpreted using a "thematic content analysis" approach (Green & Thorogood, 2004). Thematic analysis has been suggested to be a flexible and useful research tool that can provide a rich and complex account of the data. It is a descriptive and interpretative qualitative approach (Vaismoradi, Turunen, & Bondas, 2013) defined as a "constructionist method, which examines the ways in which events, realities, meanings, experiences and so on are the effects of a range of discourses operating within society" (Braun & Clarke, 2006, p. 81).

In the initial phase of data analysis, all of the translated text data were read and analyzed by four independent coders (two Koreans, one European, and one American) in order to avoid cross-contamination and ensure inter-coder reliability. The authors then came together to share and discuss their identified codes and began to develop a potential list of themes across the data. This list was reviewed independently by the authors and themes were combined, separated, or discarded, and then rechecked to determine if there were any additional themes that may have been overlooked. The authors then came back together to discuss their revised coding and themes. This iterative process continued until no further overlapping or additional themes were found, thus resulting in strong coder agreement. Finally, based on the themes identified, the research team further discussed their interpretations of the findings within the framework of the study exploring the Korean audience's narrative and perception of the issue of authenticity in the reality show. Potential disagreements between the authors were resolved during this process by consulting the text data and coder notes.

Findings

In order to explore the authenticity of the program from the viewer's perspective, we first analyzed the dual structure of the tourism setting of the program, which constructs the postmodern touristic context of the gaze, and then discussed four identified themes that emerged from the audience's online discourse regarding the authenticity of the program.

Dual structure of tourism setting in Law of the Jungle

The plot of each episode begins with scenes that show how a group of Korean celebrities survive on their own in remote areas of the world. They call themselves the "Byung-Man" tribe, named after Byung-Man Kim, the chief explorer in all the episodes of the program. Byung-Man and his crew have to live in the wild for about a week with minimal supplies, and then visit the territory of a tribe represented as untouched by civilization. In the village, they wear traditional tribal clothing and eat and sleep together to enjoy a primitive lifestyle. While the general format of the show is similar to some well-known survival television programs, such as *Man vs. Wild* in the United Kingdom (UK) and *Survivorman* in Canada, there is a difference with regard to composition of the group; the participants in the Korean reality show are mostly selected not because of their survival skills but because of their heavy media exposure as actors, comedians, or famous athletes.

The fact that all members are celebrities and complete novices in the jungle or similar non-urban areas, supports the main theme of the program and the appropriateness of the visiting area: it is not actually a documentary on how to survive but a show of performing survival in the jungle, and the places they visit in the program would be safe areas for ordinary people. The "staged-ness" of the actors' exploration is manifested in the lenient minimal supplies and the actors' interaction with the native tribes. Even though finding food poses the main challenge in all episodes, the actors are sometimes provided with Korean food rather than the unique ethnic food of the natives. A major focus of interaction with the natives is on imitating their exotic way of life for fun rather than for learning the natives' survival skills. For the audience, this program content dilutes the meaning of watching challenges for survival and encourages another way of viewing virtual travel. The audiovisual texts also contribute to the touristic appeal of the program.

Every episode begins with the visual text, "This program contains product placement". This message indicates the obvious presence of commercials on the screen even though the uncivilized jungle should be highlighted in all the images throughout the show. Yet this message is tempered by an audio track that follows the first shot:

> In the jungle, there are primitive tribes, untouched nature and survival instinct. And the law of the jungle. The Byung-Man tribe challenges the law. With experiences of their own triumphs and disasters, the conflicts and reconciliation among them in extreme situations of life and death, and their tears, here we go! (The first narration presented in the first episode)

The narrator of the program then turns the viewers' attention to touristic representations by emphasizing exotic features of the jungle that include the wildlife landscape, the natives, and survival instinct. The participants are depicted as challengers to the unknown jungle. For viewers, this juxtaposition of human and non-human factors provokes an expansion of the range of toured objects; not only are the locals, the foreign culture and their untouched natural environment described as the objects to be gazed upon, but also the emotions and interactions of the celebrities. This broadened spectrum indicates that the actors in the program take a double-edged role in relation to the viewer: they are both the toured objects and the tourists through the spectator's touristic gaze.

Viewers gazing at both the representation of the world "out there" and the Korean actors as toured objects might be explained by the voyeuristic nature of reality television. As Strain (2003, p. 3) stated, cinema and television viewing present spectators with an illusion of de-mediation, despite their mediated experiences, that "produces the illusion that certain types of experience can strip away the typical mediations that intervene in the experience of reality, or that certain technologies, through their radiant powers of representation, can somehow reverse the postmodern eclipse of authenticity". Thus, while the program *Law of the Jungle* presents itself as a tour by displaying the actors as tourists experiencing an exotic land, the viewer simultaneously engages as a tourist via the screen when he or she gazes upon the actors moving around the jungle. The act of seeing in the form of virtual travel blurs the distinction between the exotic others and the actors by considering them all as toured objects.

Not only that, but the two positions of the viewer's gaze of "being there" and "watching from a remote location" beyond the screen present a dual position for producers:

> PD: Thanks for coming here. There's a person nearby who can teach us how to survive in the jungle. His name is Bob.
> Actor1: Where is he? Should we call him?
> All actors: Bob!
> Narrator: Bob has been living in the jungle for 30 years. He seems to share his know-how with the Byung-Man tribe. (Narration 1, Audio text retrieved from the seventh episode)

> Actor 1: Oh, it's raining!
> Producer 1: Camera! Camera!
> Producer2: We should wrap up before getting heavier rain. It'd been nice just before it rained, though.
> Actor 2: The weather here is very mixed. Rain and sunshine.
> Narrator: Recordings stopped due to unexpected rain. ... The weather here is very unpredictable. (Narration 2, Audio text retrieved from the seventh episode)

In Narration 1 and Figure 1, producers did not show up on screen, but their comments confirm their role as guides for the celebrities to enter the jungle. The presence of producers becomes more visible in the second Narration 2 and Figure 2: Several producers carrying recording equipment were captured on screen when they delivered the participants a message of the suspension of recording due to a spell of rain.

The different narrative structure of the two excerpts extended the range of gazing to the presence of television cameras. In the former case, viewers enjoy a panoptic mode of observation in line with that of producers who gaze back at the actors in the recording setting. This mode of looking is broken by a spontaneous occurrence as shown in the second case. Unexpected rain caused the

Figure 1. Participants listening to the producer's direction in *Law of the Jungle*. Reproduced with permission from Seoul Broadcasting System.

Figure 2. Stop shooting shown in Episode 7 of *Law of the Jungle*. Reproduced with permission from Seoul Broadcasting System.

producers to show up on screen without breaking the aura of reality, and in turn, they exposed their back along with the actors to spectators. For audiences, the yelling of "Camera! Camera!" unconsciously transformed the producers into toured objects. The exposure of cameras and producers paradoxically increases the reality of the show by providing voyeuristic enjoyment observing the unplanned situation while revealing that viewers are really in a media-mediated situation. This spontaneity allows the spectator to enjoy the hidden through the planned camera view by collapsing the distance between the viewer and the toured objects (i.e. jungle, actors, and producers).

As investigated, the voyeuristic appeal of gazing upon the producers, the locals, and the participants manifests the dual structure of the tourism setting rooted in *Law of the Jungle* (Figure 3). The touristic aspects of the program become obvious through the characteristics of the participants, the civilized and untrained celebrities in the wild terrain, and the mixture of the audiovisual texts that inform viewers both about branded products and the exotic appeal of the site. These conditions lead viewers to interpret the actors as tourists in staged tourism settings. The mediated gaze of viewers upon the television screen transforms the role of the actors from tourist to the toured, and the role of producers from gazer to another type of participant. This dual structure of the tourism setting, in the jungle and on screen, contributes to the sophisticated but conflicted discourse on the authenticity of the program in the venue of virtual travel.

The viewers' discourse on authenticity

The data analysis of the viewers' online discourse indicated that spectators interpret the authenticity of the program from three distinct perspectives: objective, subjective, and postmodern. Within these perspectives, four corresponding themes emerged: (1) authenticity is contingent upon a perceived reality of toured objects within the frame of the camera; (2) authenticity is contingent upon a perceived reality of toured objects beyond the camera's frame; (3) authenticity connects with viewers' own perspectives; and (4) it is meaningless to seek

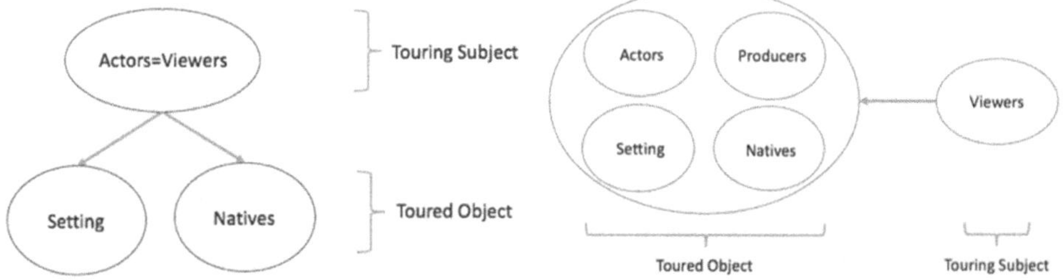

Figure 3. Dual structure of tourism setting in *Law of the Jungle*.

authenticity in the media. While the first two themes reflect the object-oriented approach to authenticity, the third and fourth themes are related to the subjective and postmodern perspectives respectively.

Theme 1. Authenticity is contingent upon a perceived reality of toured objects within the frame of camera (object-centered approach 1)

The most commonly mentioned narratives regarding the authenticity of the program were those that reflected the viewers' perceived reality of the images in the program. Two distinctive meanings of authenticity emerged: (1) authenticity as the untouched, genuine features of toured objects; and (2) authenticity in the function of the setting and cast members.

Authenticity as the untouched, genuine features of toured objects. Numerous spectators argued that the authenticity of the program appeared to have deteriorated, given that the show provided a blemished view of the scenes that the audience sees. The evidence of this was prevalent in their overarching observations of the cast members, settings, and natives. The actors were mainly considered impure intruders into the virtual travel site. Although cast members supposedly showed only their natural features in front of cameras, they "always looked neat and clean" (ID: Babmukkosalja) with "their face made up nice" (ID: suwanee08). The incongruity of this beautification was more pervasive throughout the survival settings that included hunting, eating, and sleeping in the jungle. Spectators pointed out that actors were "hunting animals every time they go out, but even natives take several days to do so" (ID: phat), and that they were "not suffering from insect bites despite sleeping in a banana leaf shelter" (ID: dreamboy). These observations led viewers to conclude that the celebrities "were not doing things to actually survive at an experience level" (ID: Solu****). The ideal landscape on the screen was implicitly portrayed as savage, and the viewers' critique of the wild reflected their desire to see the jungle as undisturbed by civilized actors and producers.

The viewers' reading bifurcated the civilized producers and actors, and the primitive and untouched jungle, and this binary approach was echoed in the critique of the locals . Viewers pointed out that natives were "usually holding some important events in time when different-looking outsiders came into the village with recording equipment" (ID: Andrew). The representation of the locals as untouched was further weakened by the natives' appearance. In the audiences' voice: "[The natives'] gold teeth gleamed with pure white teeth" (ID: Oops) and "some put on a white and clean hat that looked like a shooting prop"(ID: LonglivePark).

Authenticity in the function of the setting and cast members. A number of spectators argued the reality of the show as a negotiable thing between the manipulated and the untouched. They felt that the authenticity of their virtual travel in the camera frame was not equivalent to the objective truth of all the scenes. In this context, many of the hardships endured by cast members were considered major components, confirming the authenticity of the program: "The authenticity of the program is fully proven as [actors] have been suffering hardships including starving and getting hurt in the jungle" (ID: VitaminC). The degree of safety provided viewers with another layer of the meaning of the real:

> *Law of the Jungle* is real in so far as the minimum safety standards that prevents [actors] from suffering any possible accidents and casualties. Isn't it a prerequisite to accompany the medical team, local guides and experts when directing such programs? Also, isn't it understandable for the director to liaise with the local tribe before they arrive? ... (ID: vict****)

The social justification that the intentionally staged setting only sought to protect participants from injury was further interpreted as a need to consolidate the authenticity of the program through their hard work, such as "claiming a coconut tree with their bare hands, constructing boats and houses, and making a fire" (ID: Bbor****).

Furthermore, viewers argued that, even though producers might partially create a fabricated scene, the fact that participants actually went to a remote place and met the exotic others protected the authenticity of the program. Given the limited approach for viewers, some spectators claimed that "the program is real because [actors] truly went to the jungle" (ID: hyunhee) and "the program is authentic in that we can see the exotic natives and natural environment that we can't easily have access to in reality and that the Byung-Man tribe actually experienced their way of life" (ID: goyo****).

Theme 2. Authenticity is contingent upon a perceived reality of toured objects beyond the camera's frame (object-centered approach 2)

Exposure to external information about the local people, cast members, and producers expanded the scope of perceived authenticity of the program. Two distinctive meanings concerning authenticity emerged: (1) both the behaviors of toured objects on and off camera should be consistent to guarantee the authenticity of the show; and (2) the toured objects on the screen and their behavior off camera are located in different dimensions of reality.

Authenticity as consistency within the frame of a camera. Some viewers argued that the authenticity of the program was diminished by the discrepancy between the behavior of toured objects on and off camera. Frequently mentioned evidence was that the Waorani clan had introduced and presented themselves and their traditional lives on Facebook to appeal to prospective visitors. Audiences felt that this tourist-friendly destination was masquerading as wild in order to appeal to spectators as an ideal location for virtually exploring mystifying objects. To support this, some spectators listed photos of "Waorani people going to work by boat in indigo color skinny jeans and nice makeup" (ID: seven****). Such images of the locals apart from the show's camera made audiences consider "the natives as inauthentic who work part-time and get paid for participating in the program" (ID: Upsoo).

This type of criticism of the oxymoronic features of toured objects echoed in observations of the film crew. Actors and producers admitted that they took time off in the same way that they did in South Korea: sleeping in a hotel and drinking lots of beer. Their "townie-like" behavior off camera lead many viewers to be suspicious of the authenticity of the program due to its lack of continuity: "The major problem is that the program didn't air the scenes about the actors actually sleeping in hotels. I naively believed that they'd really slept in the jungle" (ID: let'smile). For some spectators, this leisurely break functioned heavily as a critical point to negate the authenticity of entire scenes: "[producers] tried to find ways to deceive and make fools of us" (ID: circle). And the experience of participants was reduced to "a cheap packaged tour in which cast members indulged in tired and hungry play-acting in front of the camera" (ID: nestwon).

Authenticity as inconsistency beyond the frame of the camera. Many audiences, however, rebutted the critical interpretation of the discrepancy, based on their careful reading of and listening to the narratives of the program and their sympathy for the participants. They argued that the narrative of the program offered viewers enough clues, so it was inattentive viewers, not inconsistency, that gave rise to the inauthenticity of the show:

> I anticipated that actors were supposed to stay in decent hotels when moving between destinations. Watching the interview of participants, I didn't felt that the program did intentionally conceal or hide the fact. ... Isn't it a natural and inevitable process? ... (ID: blackRussian)

The primary responsibility of viewers for their inattentive interpretation served as a buffer against suspicion about the authenticity of the scenes presented on camera. Furthermore, several of those who posted comments argued that the intermission intensified the reality of the show with respect to the long shooting period. Their previous wildlife expeditions supported this idea:

> The actors have to film for a long time period so they deserve to stay at a good hotel during their free time. ... Isn't it reasonable for the tired travelers? In my case, while I usually stay at a guesthouse when traveling abroad, I intentionally stay at good hotels the day before when I'm planning to travel to rural places. (ID: chocobanana)

Some viewers interpreted the inconsistency as an essential element for the urbanites, given that their physical weakness in the jungle environment required suitable relaxation:

> Do you think it's possible to dwell in the jungle for almost a month? No way. You will get totally exhausted even after having five days of training. It's almost impossible to stay that long in the jungle unless you are a native . ... you should understand the circumstances and reality of the site. ... (ID: corr)

Thus, taking a rest as ordinary people did was reasonable, and this commonsense action seemed to support the authenticity of the show. In this context, maintaining consistency on and off camera became ideal, not real.

Theme 3. Authenticity has a connection with viewers' own perspectives (subject-centered approach)

Compared with the aforesaid themes, other viewers focused on their significant role as touring subjects. For them, the perceived authenticity of the program mainly based on toured objects was meaningless because (1) the "real" jungle does not exist in contemporary society, and (2) authenticity is essentially interpretive.

The loss of the real jungle. Several of those who posted comments denied the existence of the "untouched". They argued that modernization as a worldwide phenomenon inevitably led the locals and the natural environment in the wild to resemble, to some extent, those living in the non-wilderness: "Even people of the tundra in Siberia are influenced by modern culture. They ride helicopters and use the Soviet language, and fire guns" (ID: nigimijo) and "frankly speaking, there are no natives who still live in a primitive and uncivilized place" (ID: stanleykubrick). Instead of criticizing the manipulation by producers, viewers pointed out that the current features of the destination with its civilized figure seemed natural, given the unavoidable submergence of wild areas in modernization:

> Nowadays, 100% wild and remote places don't exist anymore around the world. All places are to some extent civilized and influenced by modern culture. The natives who appear in the program may be the ones who live in the jungle while earning tourism profits. It is true that actors explored the jungle but for fun. (ID: namusarang)

This individual neutralized the debate on the authenticity of the toured object. For the viewer, there was no contradiction or conflict between the scenes with natives who were "civilized" and lived off tourism revenue and those with cast members who "explored the jungle for fun". This resolution was in line with that of others who accepted the modern way of life of the indigenous because they had seen it in a documentary program:

> It is a widely recognized fact that the natives of the jungle, as portrayed in the media, are the ones who actually live in the city. Even when I saw the *Tears of the Amazon* [a documentary program], natives were shown to have been embracing civilization. I guess the actors of *Law of the Jungle* met those kind of tribes. (ID: T.T)

Descriptions of the locals as modernized beings manifested the viewer's approach to the authenticity of the program: it went beyond the issue of fabrication of the toured objects as tourist attractions shaped by the camera. The real jungle in a fixed frame of the untouched is only a mythical interpretation of the viewer.

Authenticity as interpretive framework. Some viewers were in quest of the real hence their practice of seeing the program beyond the screen. They argued that the occurrence within the camera frame was a virtual travel medium for gazing at their own emotional responses and feelings: for viewers, "Even though the program is filmed with the director's intention to pretend everything is real and authentic, it is real for viewers in that they actually cry and laugh while watching the program. That's the reality" (ID: gnayungbaram). In this context, the act of watching enabled the viewer to realize the habit of looking at others from a distance and thus reducing locals to objects to be toured:

> I am ashamed of myself watching the broadcast with curiosity when the people of primitive tribes are shown. They seem like a people from a different world. I'm not sure what we can do for their survival and how to really understand the meaning of their lives in the jungle. It is somewhat uncomfortable to see the people of tribes, who had once solely relied on hunting for their living, eating meats provided by tourists who only enjoy hunting for fun.(ID: Fabian)

In this quotation, the viewer interpreted the manipulated locals as a symbol of commoditization based on their lifestyle changing from living as hunters to one of toured objects. The viewer realized that his or her habitual gazing at the representation of exotic others was rooted in curiosity. This self-reflective interpretation led the viewer to feel ashamed and to rethink how to sincerely understand others.

Another type of response concerning the viewers' interpretation was that they projected their own desires onto the screen. Thus, their main concern was not seeking authenticity of toured objects but gaining a vicarious thrill out of watching cast members wander through the jungle:

> I'm not watching [the program] to see the real and authentic. I watch it because it gives a vicarious satisfaction to me, who is getting tired of living in a desolate city. Ordinary people, like me, can't go into the jungle and have such experiences. ...Adventure is always interesting for me. (ID: Goldongi)

Theme 4. It is meaningless to seek authenticity through the real in the media (postmodern approach)

The data analysis indicated that a significant number of viewers were sceptical about the existence of authenticity in media travel. They argued that it is unsubstantiated to seek the authenticity of the program by distinguishing the real and the fake, because (1) any form of recording cannot be real, and (2) reality travel programs are intended to deliver fun, not the truth.

The media without the real. Many viewers pointed out that any form of media was incapable of representing objects without the interference of producers. They argued that documentary films and news also use manipulation, so their "packaging" led to "there is nothing that matches 100% with reality on the TV screen" (ID: scotbrain). In this context, viewers considered the image captured by the camera as an "illusion":

> Broadcasting is a huge commodity that elaborately creates fabricated illusions. News also require filtering processes and entertainers and broadcasters turn out to be artificial beings who don't have a sense of self. I enjoy television programs and fall for their lies simply because they have nothing to do with me. There is no such thing as sincerity in the context of broadcasting. It's only a product. As for this program, it will be just fine if viewers understand this situational context and don't take it seriously. (ID: HacDonald)

In this context, the image of moving individuals on the camera was not interpreted as a lived experience of the filmed objects but rather as a socially constructed representation of the experience, and in turn, even the

Korean cast members became not only "artificial beings" but also "products" or "commodities" to be consumed. An indelible line between viewers as human gazers, and actors as imaginary beings to be gazed upon in the media led viewers to disregard the thinking and judgment of the moving images. This interpretation was echoed by other viewers who argued that the reality of the program was located in a different dimension to that of the viewer. They asserted that "Byung-Man Kim is simply filming the program *Law of the Jungle* and not actually surviving in the jungle" (ID: SajabirangAkira), "even the breathing is a lie in the broadcast" (ID: Ifit'safate), and "A reality entertainment program is a program that is directed so that it looks as real as it can. It's impossible to produce such programs without having prepared scenarios and without directing" (ID: Kwongadin).

Fun beyond the reality of the program. Given the indifference toward the images on the screen, many viewers focused on the goal of the travel show – to entertain the audience – and argued its ascendancy over the authenticity of the internal context of the program. This was manifested through their description of fun that lent legitimacy to the fabrication of the program:

> It's not reasonable to simply blame the program for its manipulated settings. Given that the program is being directed in poor conditions within a limited time frame, it is understandable for producers to sometimes lose prepared animals for the show. How can you expect 100% reality from the program? You should bear in mind that [this] is an entertainment program for fun. (ID: pyb2****)

The physical constraints of filming and the manipulation of scenes were considered bearable because they respond to the needs of viewers who do not seek authenticity in entertainment:

> I understand the necessity of manipulation, to some extent, of an "entertainment" program for fun. The program will indeed be more attractive and fun when a variety of animals show up. Frankly speaking, who would be willing to watch the program if weren't fun at all? LOL. (ID: cute****)

The viewer pointed out other viewers' paradoxical attitude: although they pursued the authenticity of the program, what arrested their attention in reality was the degree of fun. In a similar vein, this type of viewer was commonly sarcastic about those who raised the issue of authenticity of the program:

> Apart from authenticity, the program is an entertainment variety show, not a documentary. If you wish to see reality with no fun factors, you shouldn't watch this. ...(ID: auau****)

> A lot more idiots than I thought. The participants are entertainers who earn several [thousands of dollars] per episode. ...The actors are playing in the scenario that the director prepared and developed. (ID: gold****)

As shown in the quotations, viewers who rated fun as the leading criterion contrasted themselves with those who neglected to acknowledge the quality of the entertainment show. Drawing a broad line between documentary films and entertainment variety shows and between traveler in reality and celebrities in the media, viewers emphasized the distinct characteristics of the show: it always deployed the gazed objects through the producer's frame in order to create entertainment, without any concern for the authenticity of what was produced.

Conclusion and discussion

This study has investigated how Korean audiences perceived the authenticity of *Law of the Jungle* through their practice of seeing. The analysis of the program narrative identified a dual structure (i.e. distance and immersion) of the tourism setting that enabled viewers to expand the range of toured objects from the locals and the physical environment to the producers and the actors. The online discourse analysis revealed four distinctive themes in viewers' interpretations of the authenticity of the travel show: (1) authenticity related to a perceived genuineness of toured objects (i.e. object-centered approach); (2) that of authenticity beyond the camera's frame (i.e. object-related approach); (3) authenticity as the viewer's own subjective interpretation (i.e. subject-related approach); and (4) the meaningless of authenticity in the media (i.e. postmodern approach).

Numerous viewers appeared to seek authenticity in the show from the actual toured objects such as the uncivilized natural environment and the natives. This object-centered authenticity was identified with four specific meanings of the perceived reality of the scenes: authenticity as the actual features of cast members, natives, and setting; authenticity as the contextual reality in the function of the setting and cast members; authenticity as consistency of the features of the locals, producers, and actors both on and off camera; and authenticity as inconsistency of the features of actors between on and off camera. These meanings appear congruent with the traditional tourism literature in which objectivists anchored in modernist paradigms (i.e. positivism and empiricism) posit that authenticity of tourist experience is largely determined by the perceived authenticity of toured objects (see for example Boorstin, 1964; MacCannell, 1973). A remarkable finding is that viewers still argued that the authenticity of the program was preserved due to the hard-working actors and their hardships, even while acknowledging manipulation of the locals and the setting. Thus it seems plausible that the audience seeks modernist authenticity in the flood of mass-media images, as Rose and Wood (2005) pointed out, and the dual mode of virtual travel leads them to fulfill such desires through diversifying their way of seeing.

It is also noteworthy that the online discourse about the object-centered authenticity mirrored the modern Western myth of primitives, as some viewers were somewhat unconsciously "pushing the Other back in time" to contrast it with themselves as modern beings (Fabian, 1983). This process of "othering" by non-Western viewers may have something to do with the voyeuristic nature of gazing. The symbolic use of the distanced gaze essentially exoticizes the other in order to gain a totalized knowledge of the object being gazed at (Strain, 2003). In fact, while gazing at others enables Korean viewers to perceive the authenticity of the program through an expanded range of toured objects, this act of seeing filters through a consistent description of the exotic others as those untouched by modern civilization. In line with the tourism literature on authenticity and orientalism, which reveals that non-Western societies consider modernity and Westernization interchangeably (Echtner & Prasad, 2003; Nyiri, 2006; Silver, 1993; Winter, 2007), Korean viewers' distanced gaze echoes the dominance of "Western self-subjectivities and self-positionings in the world order" (Yan & Santos, 2009, p. 297).

On the other hand, other viewers argued that the idea of subject-centered authenticity might rest on a foundation of resisting the uncritical interpretation of others. The findings suggest that many viewers denied the existence of the "real" jungle, and perceived the authenticity of the program by realizing their own habitual act of seeing that interpreted others as commodities and tourism products. Their approach to authenticity appears congruent with that of constructivists who rebut the idea that the authenticity of a toured object does not exist in a primitive or pre-modern society but rather should be understood as a negotiation process within the given social and cultural context (Bruner, 1994; Cohen, 1988). Furthermore, their realization appears consistent with the second gaze in which tourists are able to gain "a chance to glimpse the real" that is concealed by manipulation in the tourism context (MacCannell, 2001, p. 36).

One might argue that the denial of the "real" jungle is still rooted in finding the undistorted and intrinsic quality of the toured object. However, this objectivist approach is unsubstantiated, given that none of the viewers, whose comments were categorized as subject-centered authenticity, blamed the modernized features of the toured objects for the loss of the real jungle. Instead, their gaze at the toured object ended up as "vicarious fulfillment" (ID: Goldongi) or a "shame on [herself or himself]" (ID: Fabian) without perpetuating the objectivist order. Such psychological and emotional reactions may symbolize their changing position from bystander or conformist to dissenter who questions "what is seen as true" (Bruner, 2005, p. 90) – which also aligns with the second gaze (MacCannell, 2001, 2011).

Lastly, the findings suggest that numerous spectators interpreted the toured object represented in the media as "fantasy" and mainly focused on pursuing fun and entertainment. Viewers in this category deemed the pursuit of authenticity to be meaningless by pointing out the fact that what they see in the media is all inevitably manipulated and regulated. This kind of attitude is supported by the extant postmodern tourism literature that describes tourists as those who recognize manipulated tourism settings and enjoy the staged features (Eco, 1986; Lash & Urry, 1994; Urry, 1990).

Arguably, findings regarding the pursuit of entertainment were largely the result of the use of the word "fun" in the Korean online community. Here "fun" carries an accusation of those who are vehemently involved in an online argument. Considering the playful nature of this word, viewers who perceived a rational approach to authenticity as aggressive employed fun to nullify its seriousness. Although one may argue that this is just a kind of game for post-tourists in cyber space (Lash & Urry, 1994), such an approach reduces the use of fun to a shallow reaction and fails to embrace its significance in challenging the dominant discourse. That is, the fun-oriented statements were not only playful for the viewers but also enabled them to distance themselves from the Western-centric discourse on authenticity. Therefore, we would rather approach the viewers' quest for fun as an opportunity for disregarding authenticity as a cultural ideology that directs their virtual travel experience via the plethora of representations of tourism landscapes shaped by the tourist gaze.

Despite the focus of this study on investigating the perceived authenticity among Korean audiences through a dual mode of gaze (i.e. the tourist gaze and the second gaze) in the postmodern tourism context (i.e. watching from a distance and being there simultaneously), several limitations must be taken into account. First, all viewers were approached as virtual travelers regardless of their own subjectivities. Although the tourism literature notes that viewers of mass media are able to have touristic experiences without physical movement by blurring the boundary between mundane life and tourism (Lash & Urry, 1994), this does not mean that all viewers perceive their virtual travel as a touristic experience. Thus, further examination is required to investigate those who perceive their media travel as a tourist experience in order to understand virtual tourism from the tourist's perspective. Another limitation is that the transferability of our findings to other reality travel programs should be cautiously examined due to the unique starting point for online discourse. The exposure of manipulation by an insider directly stimulated viewers to suspect a possible manipulation of what they were seeing and to discuss the terms authenticity, manipulation, and fabrication without external compulsion. This spontaneous context significantly influenced spectators' discourse. Additionally, it should be noted that the South Korean viewer's interpretation of the representation of authentic others might not be completely applicable to that of other societies. For example, Germany, South Korea, Nigeria, and China have shown idiosyncratic patterns of social and cultural change ranging from modernization to postmodernization (Inglehart, 1997). Thus, for South Koreans, specific modes and patterns that separate them from the toured object may be different from those of other non-Western societies as well as from those of Western societies. Finally, at the methodological level, the selection of online news forums as a data source represents a limitation of this research given that it was through the use of Facebook that Kim unintentionally led viewers to discuss the authenticity of the program.

Empirical studies investigating perceived authenticity in the context of tourism consumption are scarce in the extant tourism literature (Ramkissoon & Uysal, 2010). The current study provides marketing companies and service providers with a useful example of a practical application of authenticity. Specifically, viewers' arguments and comments could be unpacked and then articulated so as to build advanced strategies to meet the expectations, anticipation, and desires of postmodern, international tourists who are immersed on an ongoing basis in a plethora of images of tourism destinations via television viewing, and as such have developed a tolerance toward (re)negotiating their viewpoints through communicating and sharing information on the Internet. Furthermore, viewers' behavior as "armchair tourists" encourages tourism practitioners to pay closer attention to the crucial role of media technologies in eroding the boundaries between everyday life and extraordinary events. Our findings suggest that the increasing access and exposure to media messages is not only a phenomenon that calls for ongoing academic investigation, but also represents the "reality" of contemporary traveling (be it actual or virtual travel).

Disclosure statement

No potential conflict of interest was reported by the authors.

References

Zhihong, B. (2007). Ethnic identities under the tourist gaze. *Asian Ethnicity, 8*(3), 245–259. doi:10.1080/14631360701595007

Baudrillard, J. (1983). *Simulations* (pp. 1–79). New York, NY: Semiotext(e).

Beeton, S. (2005). *Film-induced tourism*. Clevedon: Channel View Publications.

Beeton, S. (2006). Understanding film-induced tourism. *Tourism Analysis, 11*(3), 181–188. doi:10.3727/108354206778689808

Boorstin, O. J. (1964). *The image: A guide to pseudo-events in America*. New York, NY: Harper and Row.

Braun, V., & Clarke, V. (2006). Using thematic analysis in psychology. *Qualitative Research in Psychology, 3*, 77–101. doi:10.1191/1478088706qp063oa

Brown, W., & Singhal, A. (1993). Media and prosocial messages. *Journal of Popular Film and Television, 21*(3), 92–99. doi:10.1080/01956051.1993.9943977

Bruner, E. (1989). On cannibals, tourists, and ethnographer so. *Cultural Anthropology, 4*, 438–445. doi:10.1525/can.1989.4.4.02a00070

Bruner, E. (1994). Abraham Lincoln as authentic reproduction: A critique of postmodernism. *American Anthropologist, 96*, 397–415. doi:10.1525/aa.1994.96.issue-2

Bruner, E. (2005). *Culture on tour: Ethnographies of travel*. Chicago: University of Chicago Press.

Buzard, J. (1993). *The beaten track: European tourism, literature, and the ways to culture. 1800–1918*. Oxford: Clarendon.

Chang, H. C., & Holt, G. R. (1991). More than relationship: Chinese interaction and the principle of Guanshi. *Communication Quarterly, 39*, 251–271. doi:10.1080/01463379109369802

Cloke, P., & Perkins, H. C. (1998). Cracking the canyon with the awesome foursome: Representations of adventure tourism in New Zealand. *Environment and Planning D: Society and Space, 16*, 185–218. doi:10.1068/d160185

Cohen, E. (1988). Authenticity and commoditization in tourism. *Annals of Tourism Research, 15*, 371–386. doi:10.1016/0160-7383(88)90028-X

Cohen, E. (2008). The changing faces of contemporary tourism. *Society, 45*(4), 330–333. doi:10.1007/s12115-008-9108-2

Cohen, E., Nir, Y., & Almagor, U. (1992). Stranger-local interaction in photography. *Annals of Tourism Research, 19*, 213–233. doi:10.1016/0160-7383(92)90078-4

Crary, J. (1990). *Techniques of the observer: On vision and modernity in the nineteenth century*. Cambridge, MA: MIT Press.

Crouch, D. (2000). Places around us: Embodied lay geographies in leisure and tourism. *Leisure Studies, 19*, 63–76. doi:10.1080/026143600374752

Crouch, D., Aronsson, L., & Wahlstrom, L. (2001). Tourist encounters. *Tourist Studies, 1*(2), 253–270. doi:10.1177/146879760100100303

Culler, J. (1981). *The pursuit of signs*. Ithaca, NY: Cornell University Press.

Echtner, C., & Prasad, P. (2003). The context of Third World tourism marketing. *Annals of Tourism Research, 30*, 660–682. doi:10.1016/S0160-7383(03)00045-8

Eco, U. (1986). *Travels in hyper reality*. San Diego, CA: Harcourt Brace Jovanovich.

Edensor, T. (2001). Performing tourism, staging tourism: (Re) producing tourist space and practice. *Tourist Studies, 1*(1), 59–82. doi:10.1177/146879760100100104

Everett, S. (2009). Beyond the visual gaze? The pursuit of an embodied experience through food tourism. *Tourist Studies, 8*(3), 337–358. doi:10.1177/1468797608100594

Fabian, J. (1983). *Time and the other: How anthropology makes its object*. New York, NY: Columbia.

Fiske, J., & Hartley, J. (1978). *Reading television*. London: Methuen.

Foucault, M. (1963). *The birth of the clinic*. New York, NY: Random House.

Franklin, A., & Crang, M. (2001). The trouble with tourism and travel theory. *Tourist Studies, 1*(1), 5–22. doi:10.1177/146879760100100101

Friedberg, A. (1993). *Window shopping: Cinema and the postmodern*. Berkeley: University of California Press.

Fürsich, E. (2002). Packaging culture: The potential and limitations of travel programs on global television. *Communication Quarterly, 50*(2), 204–226. doi:10.1080/01463370209385657

Goffman, E. (1959). *The presentation of self in everyday life*. New York, NY: Doubleday Anchor.

Green, J. C., & Thorogood, N. (2004). Analyzing qualitative data. In D. Silverman (Ed.), *Qualitative methods for health research* (pp. 173–200). London: Sage.

Hartley, J. (1999). *Uses of television*. London: Routledge.

Harvey, D. (1990). Between space and time: Reflections on the geographical imagination. *Annals of the Association of American Geographers, 80*, 418–434. doi:10.1111/j.1467-8306.1990.tb00305.x

Hu, S. H. (2013, February 12). Byungman Kim cried for feeling like a criminal. *Newsen*. Retrieved from http://media.daum.net/entertain/newsview?newsid=20130212163110371&RIGHT_ENTER=R12

Inglehart, R. (1997). *Modernization and postmodernization: Cultural, economic, and political change in 43 societies*. Princeton: Princeton University Press.

Jay, M. (1993). *Downcast eyes: The denigration of vision in twentieth-century French thought*. Berkeley: University of California Press.

Jordan, F., & Aitchison, C. (2008). Tourism and the sexualisation of the gaze: Solo female tourists' experiences of gendered power, surveillance and embodiment. *Leisure Studies, 27*(4), 329–349. doi:10.1080/02614360802125080

Kirschenblatt-Gimblett, B. (1998). *Destination culture: Tourism, museums, and heritage*. Berkeley: University of California Press.

Lacan, J. (1981). *The four fundamental concepts of psychoanalysis*. New York: W.W. Norton.

Larsen, J. (2014). The tourist gaze 1,0, 2.0, and 3.0. In A. Lew, C. Hall, & A. Williams (Eds.), *The Willy Blackwell companion to tourism* (pp. 304–313). Oxford: John Wiley & Sons.

Lash, S., & Urry, J. (1994). *Economics of signs and space*. London: Sage.

Lee, S. Y. (2001). Tourist gaze: Universal concept? *Tourism, Culture & Communication, 3*, 93–99. doi:10.3727/109830401108750724

MacCannell, D. (1973). Staged authenticity: Arrangements of social space in tourist settings. *American Journal of Sociology, 79*, 589–603. doi:10.1086/225585

MacCannell, D. (1989). *The tourist: A new theory of the leisure class*. New York, NY: Schocken Books.

MacCannell, D. (2001). Tourist agency. *Tourist Studies, 1*(1), 23–37. doi:10.1177/146879760100100102

MacCannell, D. (2011). *The ethics of sightseeing*. Berkeley: University of California Press.

Maoz, D. (2006). The mutual gaze. *Annals of Tourism Research, 33*(1), 221–239. doi:10.1016/j.annals.2005.10.010

Markwick, M. (2001). Postcards from malta image, consumption, context. *Annals of Tourism Research, 28*(2), 417–438. doi:10.1016/S0160-7383(00)00049-9

Molina, A., & Esteban, A. (2006). Tourism brochures: Image and usefulness. *Annals of Tourism Research, 33*(4), 1036–1056. doi:10.1016/j.annals.2006.05.003

Nyiri, P. (2006). *Scenic spots: Chinese tourism, the state, and cultural authority*. Washington: University of Washington Press.

Patino, A., Kaltcheva, V. D., & Smith, M. F. (2011). The appeal of reality television for teen and pre-teen audiences. *Journal of Advertising Research, 51*, 288–297. doi:10.2501/JAR-51-1-288-297

Perkins, H. C., & Thorns, D. C. (2001). Gazing or performing? Reflections on Urry's tourist gaze in the context of contemporary experience in the Antipodes. *International Sociology, 16*(2), 185–204. doi:10.1177/0268580901016002004

Pritchard, A., & Morgan, N. J. (2000). Privileging the male gaze: Gendered tourism landscapes. *Annals of Tourism Research, 27*(4), 884–905. doi:10.1016/S0160-7383(99)00113-9

Ramkissoon, H., & Uysal, M. (2010). Testing the role of authenticity in cultural tourism consumption: A case of Mauritius. *Tourism Analysis, 15*, 571–583. doi:10.3727/108354210X12889831783396

Riley, R. W., & Van Doren, C. S. (1992). Movies as tourism promotion: A 'pull' factor in a 'push' location. *Tourism Management, 13* (3), 267–274. doi:10.1016/0261-5177(92)90098-R

Rose, R., & Wood, S. (2005). Paradox and the consumption of authenticity through reality television. *Journal of Consumer Research, 32*(2), 284–296. doi:10.1086/432238

Said, E. (1978). *Orientalism: Western representations of the Orient*. London: Routledge and Kegan Paul.

Santos, C. A. (2004). Perception and interpretation of leisure travel articles. *Leisure Sciences, 26*(4), 393–410. doi:10.1080/01490400490502462

Selby, M. (2003). *Understanding urban tourism: Image, culture and experience*. London: I.B.Tauris.

Shaffer, T. (2004). Performing backpacking: Constructing 'authenticity' every step of the way. *Text and Performance Quarterly, 24*, 139–160. doi:10.1080/1046293042000288362

Silver, I. (1993). Marketing authenticity in Third World countries. *Annals of Tourism Research, 20*, 302–318. doi:10.1016/0160-7383(93)90057-A

Strain, E. (2003). *Public places, private journeys: Ethnography, entertainment, and the tourist gaze*. New Brunswick, NJ: Rutgers University Press.

Strain, E. (1998). E. M. Poster's anti-touristic tourism and the sightseeing gaze of cinema. In C. Degli-Esposti (Ed.), *Postmodernism in the cinema* (pp. 147–166). New York, NY: Berghahn Books.

Taylor, J. (1994). *A dream of England. Landscape, photography and the tourist's imagination*. Manchester: Manchester University Press.

Taylor, J. (2001). Authenticity and sincerity in tourism. *Annals of Tourism Research, 28*, 7–26. doi:10.1016/S0160-7383(00)00004-9

Teo, P., & Yeoh, B. S. A. (2001). Negotiating global tourism: Localism as difference in Southeast Asian theme parks. In P. Teo, T. C. Chang, & K. C. Ho (Eds.), *Interconnected worlds: Tourism in Southeast Asia* (pp. 137–154). Amsterdam: Pergamon.

Uriely, N. (2005). The tourist experience: Conceptual developments. *Annals of Tourism Research, 32*(1), 199–216. doi:10.1016/j.annals.2004.07.008

Urry, J. (1990). *The tourist gaze: Leisure and travel in contemporary societies*. London: Sage.

Urry, J. (2002). *The tourist gaze* (2nd ed.). London: Sage.

Urry, J., & Larsen, J. (2011). *The tourist gaze 3.0* (3rd ed.). London: Sage.

Vaismoradi, M., Turunen, H., & Bondas, T. (2013). Content analysis and thematic analysis: Implications for conducting a qualitative descriptive study. *Nursing and Health Sciences, 15*, 398–405. doi:10.1111/nhs.12048

Wang, N. (1999). Rethinking authenticity in tourism experience. *Annals of Tourism Research, 26*, 349–370. doi:10.1016/S0160-7383(98)00103-0

Wilson, T., & Suraya, R. M. Y. (2004). The tourist gaze goes on-line. *Tourist Studies, 4*(1), 69–92. doi:10.1177/1468797604053079

Winter, T. (2007). Rethinking tourism in Asia. *Annals of Tourism Research, 34*(1), 27–44. doi:10.1016/j.annals.2006.06.004

Wollen, T. (1991). Over our shoulders: Nostalgic screen fictions for the 1980s. In J. Corner & S. Harvey (Eds.), *Enterprise and heritage* (pp. 178–193). London: Routledge.

Yan, G., & Santos, C. (2009). "China, forever": Tourism discourse and self-orientalism. *Annals of Tourism Research, 36*(2), 295–315. doi:10.1016/j.annals.2009.01.003

Zhihong, B. (2007). Ethnic identities under the tourist gaze. *Asian Ethnicity, 8*(3), 245–259. doi:10.1080/14631360701595007

Index

Note: Figures are indicated by italics and tables by bold type.